Innovation and Entrepreneurship
Choice and Challenge

Innovation and Entrepreneurship
Choice and Challenge

Otto Chui Chau LIN
Honorary Professor
The Hong Kong Polytechnic University, Hong Kong

World Scientific

NEW JERSEY • LONDON • SINGAPORE • BEIJING • SHANGHAI • HONG KONG • TAIPEI • CHENNAI • TOKYO

Published by

World Scientific Publishing Co. Pte. Ltd.

5 Toh Tuck Link, Singapore 596224

USA office: 27 Warren Street, Suite 401-402, Hackensack, NJ 07601

UK office: 57 Shelton Street, Covent Garden, London WC2H 9HE

Library of Congress Cataloging-in-Publication Data
Names: Lin, O. C. C. (Otto C. C.), 1938– author.
Title: Innovation and entrepreneurship : choice and challenge / Otto Chui Chau Lin,
The Hong Kong Polytechnic University.
Description: New Jersey : World Scientific, [2017] | Includes bibliographical references.
Identifiers: LCCN 2017026741 | ISBN 9789813146600 (hard cover)
Subjects: LCSH: Technological innovations--Economic aspects. | Entrepreneurship.
Classification: LCC HC79.T4 L5543 2017 | DDC 338/.04--dc23
LC record available at https://lccn.loc.gov/2017026741

British Library Cataloguing-in-Publication Data
A catalogue record for this book is available from the British Library.

For any available supplementary material, please visit
http://www.worldscientific.com/worldscibooks/10.1142/10195#t=suppl

Desk Editor: Tay Yu Shan

Typeset by Stallion Press
Email: enquiries@stallionpress.com

Printed in Singapore

***This book was written in dedication to my grandchildren*:**
Christopher, Markus, Jessica, Joshua, Elizabeth
(子弘, 子明, 子英, 子強, 子文)
and those who may come afterward.

Be reminded that —
"Polish brass as mirror, to straighten up posture,
Observe people as mirror, to know gain and lose,
Study history as mirror, to learn rise and fall."

以銅為鏡, 可以正衣冠;
以人為鏡, 可以明得失;
以史為鏡, 可以知興替。

A quotation from Li Shi Min (李世民 ***598-649, AD***),
King Tai-Zong of Tang Dynasty (唐太宗)

PREFACE

During the last 25 years I have traveled to many cities in different parts of the world and spoke at various conferences. The questions I heard most frequently were:

- How can I build in my region a profitable semiconductor or computer factory?
- How can my university turn thousands of scientific papers to a few commercial successes?
- How can I commercialize the findings of this excellent research?
- How can I be a millionaire before age 30?

These questions were from people of different backgrounds and responsibilities and were asked in different languages and in different ways.

The message is clear. Most people are interested in acquiring knowledge; but more importantly, in applying knowledge to create value and wealth. In short, the crux of the question is how to harvest economic benefits from knowledge.

What I have tried to tell them are these:

- Scientific research, technology development and commercialization are related but different; they should be addressed differently.
- Complexity and difficulty increase very rapidly from research to innovation to entrepreneurship.
- Entrepreneurship is a team effort, no one can do it alone.

- There is a need to build a culture of entrepreneurship, which is in the realm of soft power.

This book is an attempt to put these thoughts in its simplest form. I argue that it is of fundamental importance to build an innovation system in a region of interest so that the institutional players (government, university, institute and business) can work congenially and to facilitate the effort of individual entrepreneurs. Thus, knowledge can be brought to bear on the livelihood of the people.

I have intended to address the practical side of entrepreneurship, not the scholastic side. The five parts of the book will guide the reader from learning the basics, to the practice, to recognizing the challenges of change. Then it shows examples of best practices and role models.

The process of commercialization of research results is by no means straightforward. In this book, different models of developing innovation for companies covering a wide range of products and services are highlighted. Examples include the classical case of Fairchild Semiconductor in integrated circuits, to Cisco Systems for computer networks, TSMC for IC Foundry, Ardentec for wafer testing, Delta Electronics for parts and components, Epistar for high brightness LED, Raytheon's microwave oven, GIANT's carbon fiber bicycles, to Alibaba for e-commerce and Tencent for group chats.

The parts, however, are written quite independent of each other. Each part can be read separately. The reader can go from the beginning to the end, or, hop to any place that catches his fancy, and start over later.

- Part One illustrates the concept of innovation and entrepreneurship and the components of an effective innovation system. It should be of interest to every reader.
- Part Two provides more detail on what a national innovation system is and how an effective eco-system can be built. What should be the roles of the government, university, business and technology institutes? This part should be of special interest to the leadership circle of government, business and organizations.
- Part Three discusses the concepts of Laozi and Confucius as related to innovation, entrepreneurship and the management of change. It shows how innovation can be made more competitive and entrepreneurship can

be managed to a successful end. And, importantly, how the spirit of entrepreneurship can be nurtured. This will give readers a glimpse of oriental wisdom, which may be a surprise to many. ***Any reader can start from Part Three first.***

- Part Four examines the legacy of four of the most successful innovator groups in the world today: Silicon Valley, Hsinchu, Shenzhen and Singapore. The focus is on their formative stages.
- Part Five tells the story of seven very successful entrepreneurships in terms of the people and the company, the challenges they have encountered and the choices made. Much of the discussions in the previous four parts are reflected in these cases.

I hope the book will give different perspectives for readers to understand the salient aspects of innovation and entrepreneurship. I hope it will provide some stimulation for the younger generation to undertake the challenge of innovation and entrepreneurship, and be better prepared for the future.

ACKNOWLEDGMENT

I shall like to express my gratitude to many friends and colleagues who have assisted me in various ways: Bay Wu (吳蓓) and Zhi-Peh Wang (汪智沛) in Shenzhen; Ray Q.L. Chen (陳慶雷) in Hangzhou; Tianning Xu (徐天寧) in Suzhou; Nien-yu Chen (陳念玉) and Gloria Chang (張國儀) in Taipei; and Eric Yeung (楊國豪), Athena Leung (梁惠嫻), Adelaine Lim (林美玲) and Pauline Chau (周崇英) in Hong Kong. Special thanks are due to Alex Wai (衛炳江) and The Hong Kong Polytechnic University for extending hospitality and support during the course of this work.

I should especially thank Bay Wu's role in organizing a number of interviews and preparing notes, Pauline Chau's help in managing the manuscript piece by piece, and my daughter Ann, who gave assistance whenever I called. Finally, I must express appreciation to my wife Ada for giving me time-outs. She had thought my time after retirement are all hers; found out differently but supported the work graciously nevertheless.

CONTENTS

Part One
Back to Basics

What drives progress from an agricultural to an industrial to a knowledge economy?

- *How did physical strength become overtaken by machine power and brain-power in the economy?*
- *How were riches made? Who received the most benefit in different phases of economic development?*
- *Can knowledge be the generator, distributor and eventual equalizer of wealth?*

This book is about man and his effort to enrich his livelihood. It focuses on the elements of his endeavors: the tools, skills, methodology, mindset, social network and the environment. These fall within the realm of technology.

Hence technology is an essential element of the economy in human history. It takes different forms and functions differently at different stages of human civilization.

Chapter 1
DRIVER OF ECONOMIC GROWTH

The Agricultural Economy

The agricultural economy was prevalent in ancient times. Classical Chinese literature has documented the teachings of Confucius (551–479 BC) who stated that the king should first act with high moral standards. This would attract people, secure land and natural resources to promote the common good.[1]

People, land and natural resources were thus considered factors of production in the agricultural economy. Obviously, people with physical strength constituted the basic workforce. Land is the ground to grow crops. Natural resources — such as water — are essential to vegetation. Minerals and forestry are for the provision of better tools and methodologies.

Competitive advantage in the agricultural economy arose primarily from these factors. Wealthy landlords were usually the ones who could manipulate such factors to their advantage. On a larger scale, the richest landlord was the king. When the king felt he needed or wanted more, he waged war against another country to capture more people as labor, more land for growth and more natural resources to fulfill his ambitions.[2]

Knowing that, by and large, the total stock of population, land and natural resources does not change much over the span of a lifetime, competition among nations are usually a zero-sum game. In such games, there are clear winners and losers based on which players get more or less of a fair share.

In the agricultural economy, technology takes the form of better tools, seeds, fertilizers, more efficient skills and more effective uses of resources.

Technological knowledge is accumulated over time, recorded and transferred through generations.

One important aspect of farming life was to learn from nature by observing and studying the earth's relationship with the sun and the moon. The continuous cycle of change seen has been of great fascination to man. Thus the lunar calendar, a record of these changes, was an extremely useful guide.

In China, the lunar calendar is generally called the agriculture calendar. It provides detailed instruction on the 24 cyclic phases (*jie qi*) of the year. It forecasts with admirable accuracy when major changes in climate, humidity and temperature will occur, and, when frost and snow will begin and end. This provides guidance for farming: the suitability of doing or not doing certain things at particular times of the year. Kings and queens come and go in the course of millennia but the agriculture calendar prevails and actually improves. Interestingly enough, in America, the Farmers' Almanac fulfills similar functions based on cumulative folk knowledge and humor. It has remained on bestselling lists for two centuries.

The Industrial Economy

The invention of the steam engine by James Watt in Great Britain in 1781 marked the high point of the Industrial Revolution and brought forth the industrial economy.[3] With steam as the power source, man gradually leapt forward to the manufacturing era. Innovations in textiles, machine tools, paper-making, iron-forging, ship-building, metallurgy and chemicals have greatly enhanced the quality of life. Population, living standards and Gross Domestic Product (GDP) per capita have rapidly increased. These technological advancements placed Europe as the undisputed leader of the world's economic development. The role of technology in the new economy was abundantly demonstrated.

A new phase of the industrial economy came in the early 1880s with the generation of electric power in Europe and America. More than a genie that can provide great power, electricity has created freedom and new options and opportunities to human beings. Men no longer need to work in open fields at the mercy of sunlight. Electricity can be generated, stored and distributed to homes, offices and various other places of work at any time. This has greatly broadened the versatility and diversity of the economy and has brought new enjoyments to life. Innovations in land, marine and air transportation have further extended man's frontier. The world gradually turned into a global village.

Thus, in the industrial economy, the leading factors of production were gradually taken over by machinery, money and management. Machinery increases the productivity per unit of manpower, land and resources in all economic activities. More importantly, it can lead to new products and processes. Money is needed to build or procure machinery and to run the business. Management is important because all the production factors: labor, land, natural resources, money and machinery have to be organized and optimized for the greatest possible benefit to stakeholders.[4] Thus management science has gradually become positioned at the center of the economy and is considered an essential part of knowledge.[5]

People who have accomplishments in an industrial sector become the rich of the society. Well known examples include J. D. Rockefeller in oil, Pierre Du Pont in chemicals, Cornelius Vanderbilt in railroads, Andrew Carnegie in steel, Henry Ford in automobiles, Alfred Sloan in banking and Walt Disney in entertainment. They have all garnered personal wealth matching that of kings and queens of the olden times. The center of innovation has gradually shifted from West Europe to North America.

Technology has also created a paradigm shift in social norms. Industrial economy is the economy of scale and diversity. People with different talents can collaborate to create greater wealth than they can achieve individually. Thus economic activities in an industrial economy can be non-zero-sum games.[6] In any business process, it is desirable to seek the combination and optimization of various contributing factors to achieve the best performance. There is ample evidence that wealth can be created from none, and efficiency and effectiveness can be reached if people learn to cooperate rather than just compete.

The progress of the industrial economy and its generation of wealth were not without peril. With expanded economic development comes the depletion of natural resources: oil, minerals and water, to name a few. Industrial processes, more often than not, create toxic chemicals that may bring health hazards. Its exhaust can be poisonous and detrimental to the natural land or water environment. The emission of carbon dioxide, as harmless as it may seem in small quantities, can clutter the stratosphere and be present as heat barriers leading to global warming. It has been recognized that there are more hot days a year now than there were 20 years ago. The weather pattern in many continents has also become more erratic. It has created detrimental effects to biodiversity and the human habitat. There is a real worry that man may actually hasten the destruction of the earth if some human activities remain unchecked.[7]

The Knowledge Economy

The advent of personal computers and popularization of information and wireless technology have ushered in the era of the knowledge economy.[8] Knowledge can be quickly used to generate value, both tangible and intangible. One very dramatic change this brings to life is the speed of communication. A letter from Hong Kong to New York would generally take about one week by airmail. If transmitted by email, it will be around one second, an almost 1 million-fold reduction in time! It is little wonder that new knowledge can be utilized to create new wealth or generate new value with exceptional speed.

But the same can also be said for disasters. The Severe Acute Respiratory Syndrome (SARS) virus that originated in Asia was expected to cause health havoc in the U.S. within two months; the collapse of the stock market on the New York Wall Street on September 15, 2008 spread to all stock exchanges worldwide almost overnight. It immediately caused a tsunami of global economic declines.[9]

With the transmission of information at an explosive rate, human responses are demanded proportionally. Thus vacation is no longer an excuse for delayed action. People can become busier and get more things done even on vacation.

It may be fair to say that the knowledge economy started to emerge in the 1980s and has progressed rapidly since the start of the 21st century. The global village has created great opportunity for all.[10] People all over the world are now tightly connected by webs of technical, financial, cultural and more recently, social media networks.

It has been clearly demonstrated that the driver of economic growth has changed from physical power in the agriculture economy, to machine power in the industrial economy, and, to brainpower in the knowledge economy as shown in Figure 1. The paradigm of operation has changed from competition to win, to cooperation to create, to networking to innovate.

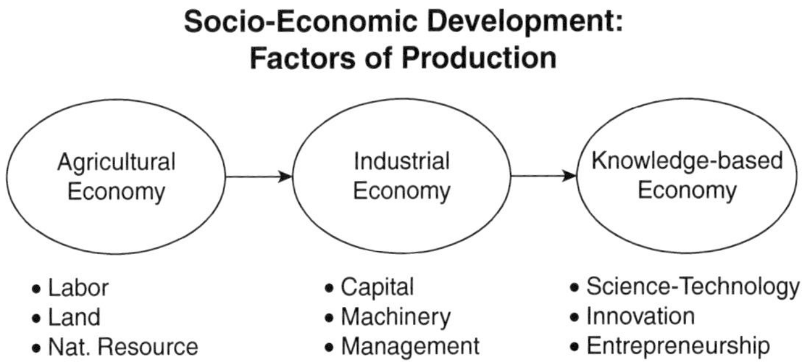

Figure 1. Factors of Economic Production

The factors of production can now be identified as scientific exploration, innovation and entrepreneurship.[11,12]

This does not mean that labor, land, natural resources, machinery, money and management, those factors instrumental in the previous economies are now unimportant. It only says that they will recede into the background in a supporting role to the development of scientific knowledge, agility in applying that knowledge and dedication to success through entrepreneurial efforts.

The heroes in the knowledge economy are entrepreneurs of exceptional dedication who can take advantage of the moment. They can be common people with humble backgrounds. Bill Gates, Steve Jobs and Mark Zuckerberg were not members of any National Academy, held no professorial honors of any sort, and did not have the privilege of owning any huge estate or powerful machinery. In fact, most did not even finish college when they were pursuing their vision. They were young adults with great curiosity and willingness to explore the unknown.[13]

They are called entrepreneurs. They are able to recognize the uniqueness of a concept, articulate its advantage, adopt and adjust a strategy, build a good team, leverage the environment and pursue their goals with exceptional diligence. They have attained success at a much higher level and in a much shorter time than leaders in the industrial economy could have dreamed of. This is the nature of the knowledge economy.

In short, the knowledge economy is one that promotes the application of knowledge to create new value, or new wealth, very rapidly. The knowledge society will provide abundant opportunities for all individuals, not just the elites. The factors of productivity in the knowledge economy will be discussed in greater detail in ensuing chapters.

The environmental problems that arose in the industrial economy continue to plague the world, especially the underdeveloped and developing nations. The carbon dioxide level in the stratosphere has accumulated over the years to an alarming point. It has led to a clear trend of global warming. Most recently, the Fourth National Climate Change Assessment (NCA4) has clearly documented that the global annually averaged surface temperature has increased by 1.8°F (1°C) over the last 115 years (1901–2016). This period is now the warmest in the history of modern civilization.[190] Global warming has caused the unusual climate change, melting of iceberg, land erosion and calamity in biodiversity. The average sea level has risen by about 7 to 8 inches since 1900, with about half of the rise occurring since 1993.

Several world conferences have been convened since 1992 under the leadership of the United Nations Framework Convention on Climate Change (UNFCCC) to map strategy for the control of carbon dioxide emissions. There were differences between developing and developed nations on the issue due to different social and political perspectives. But, all have agreed that the CO_2 level needs to be controlled substantially downward.[14,15]

In the 2015 Paris Agreement, 195 countries signed an agreement for concerted efforts to reduce the emission of greenhouse gases in 2050 to a level 26–28% below that of 2005. This would keep global warming to below 2°C. Participating members agreed to take action: cleaner power plants, more fuel-efficient cars, more green energy, eco-friendly appliances, etc. On June 1, 2017, to the dismay of member countries, U.S. President Donald Trump announced the withdrawal of the United States from the Paris Agreement.[165] This rebuke of the world effort was unusual, since the U.S., one of the top carbon emitters has played a leading role in the efforts to control global warming for the last 25 years, over three changes of presidency. Despite Trump's announcement, leading participating countries have stated that the efforts stipulated by the Paris Agreement are "irreversible." At the G20 Summit in Hamburg, Germany on July 8, 2017, the 19 member countries decided to move forward collectively on climate change without the United States.[166]

Another ominous problem is the distribution of wealth, the widening of the wealth gap. As the rich members of society are in a better position to access knowledge and become richer more easily, the poor are often left behind. Thus the divide between the rich and the poor widens. Polarization is not only serious in relation to developed and undeveloped countries but is also sharpening between people on Wall Street and those on Main Street. Those who sit on the far end of the money chain, such as the financier or the stock marketer, can manipulate the flow of currency and reap unusual profits, adding to the ills of social injustice. The September 2011 Occupy protest on Wall Street[16] created sympathetic movements in 950 cities over 82 countries.[17,18]

A 2017 Report from The Oxford Committee for Famine Relief (Oxfam) indicates that the wealth of the richest 1% people of the world equals the sum of the other 99%.[19] The net worth of the top eight richest: Bill Gates, Amancio Ortega, Warren Buffett, Carlos Slim Helu, Jeff Bezos, Mark Zuckerberg, Larry Ellison and Michael Bloomberg, totaling about US$426.2 billion, equals roughly the sum of the income of the lower 50% of the world population. All these eight gained their wealth mainly from technology businesses including semiconductors, computers, information technology, telecommunications, the internet and media, or technology-driven businesses such as electronic commerce, electronic banking and related technology services. It should be mentioned that many of the wealthiest have individually played active roles in philanthropy for the public good.

It is of interest to see the change in wealth of the world's top 100 richest people portrayed in the following chart, Figure 2.[20]

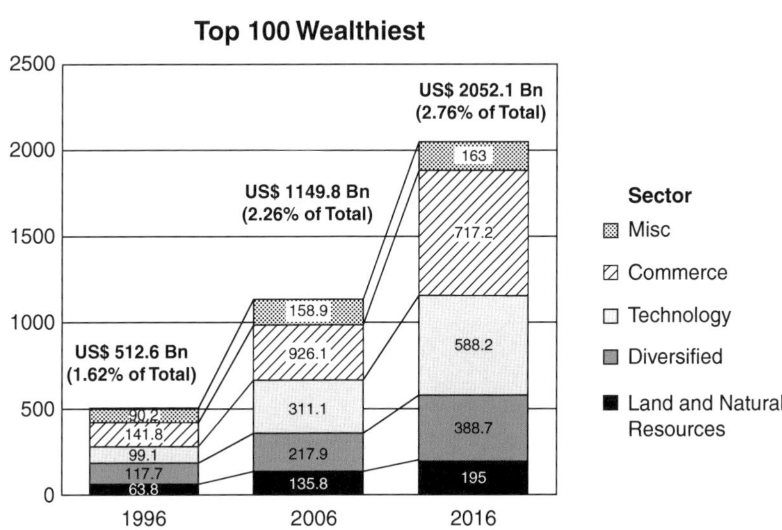

Figure 2. Growth of The Top 100 Richest Over Two Decades

Several observations can be made:

- As a group, the wealth of this group grew much faster than the GDP.
- Those focused on land and natural resources, the economic drivers of the past era, grew only mildly.
- Those focused on technology, the driver of economy, grew at a much faster rate.
- Those focused on commerce driven by technology grew at the fastest rate.

This has highlighted the role and dilemma regarding technology in economic development. Whereas technology is instrumental in fostering economic growth, it also carries hidden risks that may adversely affect social justice and environmental sustainability. How man can make technology foster economic progress and at the same time build social justice and environment sustainability remains a challenge for the future.

Chapter 2

INNOVATION

Innovation is the application of knowledge to create new wealth or new value. The term innovation is often confused with invention.[21]

Invention, the creation of knowledge through scientific research and exploration, is a technical process. The result of invention is new knowledge that generally transcends national boundaries. It can be new observations or interpretations of the nature regarding materials, processes, principles or theories. It can be in the form of publications, patents or know-how. It prizes novelty, originality, precision, sophistication and superiority.

Innovation, the application of knowledge through technology development, is an economic process. The result of innovation is the creation of new wealth or new value that adds to public goods. It focuses on efficiency and effectiveness. How novel or superior it is in certain characteristics is only secondary. But it has to provide value specific to the time and place.

Back in the 1940s, the Austrian economist J. A. Schumpeter showed that major waves of innovations such as hydropower, the steam engine, electric power, textile, chemicals, rail transportation, etc. were the major driving forces of economic growth for society and nations.[22] In the age of globalization, information travels much faster than before. Innovation in one place is rapidly copied in other parts of the world. This can shorten the wave of innovation and fuel the rapid growth of the world economy.

In the last two decades, many aspects of innovation regarding terminology, scope and relevance have attracted a wide range of interest by governments and academia.[23,24] Figure 3 is a table summarizing the salient features of innovation as I see it.

Innovation

- A process that focuses on implementation and efficiency
- Covers many forms:
  - Product/Service
  - Processing/Manufacturing
  - System integration/Continuous improvement
  - Business Model
- Requires a team of multi-disciplinary talents with entrepreneurial spirit
- Requires a supportive eco-system

Figure 3. Significance of Innovations

It should be noted that innovation comes in many forms. It can be a product that provides specific functions perceived to be of value to users, or, a service, a process, integration of processes and the continuous improvement of processes. Equally importantly, it can be a business model.

Product

Product innovation is most easily recognized by the public. It generally is a result of technology development. It has a physical presence and can be seen, touched or felt. The product's size can be large or small; the intended user can be a corporation or individual, the technology involved high or low. Numerous examples can be found in our daily life, and industrial companies have dedicated a great deal of effort to designing winning products.[25] In recent decades, the most popular technological product innovations include personal computers — desktop, laptop and handheld — mobile phones and smartphones, wireless devices, safety devices for automobiles, personal healthcare devices, etc. These are all durable products. Some come in simple packages. Most, such as smartphones, consist of multiple components, parts and sub-systems.

Product innovations have generally been made in materials, design and manufacturing. Any existing products can be improved and upgraded through these three elements.

Food and consumable products are other major categories. Technology innovations will generally take place in processing and packaging. Pharmaceutical products are yet another category of product innovation that generally involves advanced bio-medical scientific research, field tests and technology development.

Service

Service innovations cover a wide range and service firms constitute the largest sector of the most developed economies. This is also a very competitive sector.[26] The major types of service are either manufacturer-oriented or consumer-oriented. The former refers to services providing sets of data, information, instructions and consultation needed in the manufacturing of products. The latter refers to services providing relevant data, information and instruction to assist consumers in using them.

There are other services connecting the two. For example, cargo container carriers move oils, chemicals, machinery and foodstuff over continents for manufacturers, processers and consumers. Delivery by courier to customers is another classic example of service. Federal Express (FedEx), DHL and United Parcel Service (UPS) connect manufacturers and consumers and become key linkages in the commerce value chain.

In addition to big name players in the global supply chain network, there are many smaller players or agents operating in smaller areas offering more localized or personalized services. These "last mile" services are often the key element in the value chain. It is little wonder that new innovations including unmanned aerial vehicles are viewed with increased importance in the value chain. Overall, to meet rising environmental and cost constraints, supply chain innovations consisting of multi-modal services for the best efficiency and legal compliance will be constantly in demand.

Computer Software

A major service category in the knowledge economy is computer software.[27] There is a great variety of software. System software defines the architecture and algorithm of a computer and constitutes its central control. If the computer is the brain of smart machinery, then system software is the brain of the computer.

Application software provides the format and platform for various applications, such as word processing, music, video, animation, education, maps, information search and their various combinations. Development of software is different from hardware and generally requires different sets of project management skills.[28]

Cloud computing opens the world of computer infrastructure, platforms and services as a service (Infrastructure as a Service, Platform as a Service and Service as a Service, or IaaS, PaaS and Saas, respectively) to users. It has quickly become a field of fierce competition.[29] Included in this world are various web servers, corporate management programs and individual platforms such as email, Dropbox, Page, Keynote, Skype, Photoshop, Apps, etc. Most computer companies are active in this field with leaders such as Amazon, Apple, Google, Facebook and Microsoft. They compete in delivering the value that new innovations create for users.

Continued Improvement and Network Integration

System integration and continuous improvement offer innovations in the manufacturing and processing of modern products and services that customers have generally taken for granted.[30] For example, today's Apple iPhone 6 is a handheld device that fits in a suit pocket but has computing power equivalent to a Cray supercomputer of the 1980s with only about 1,000th of the size and 10,000th of the cost. This would not be possible without the concerted efforts in design innovation and system integration of Foxconn and its network of about 350 suppliers of parts and components.[152] The miniaturization of robotics in precision machinery and medical devices are other examples.

Social Media Network

Social media networks such as Facebook, LinkedIn, Twitter and YouTube are among the most intriguing innovations in the knowledge economy.[31] These networks offer opportunities for people to socialize online free of charge, in exchange for viewing advertisements as the socializing progresses. For business and industrial advertisers, this creates unprecedented opportunities to reach a wide range of potential clients transcending national boundaries. A tremendous amount of financial value has been created. Additionally, they have a lasting influence on the social and political developments of our time. The Arab Spring and ensuing democratization attempts in the Middle East in the last decade were examples.

The unexpected success of Donald Trump in the 2016 U.S. presidential campaign was, to a large extent, attributed to Trump's mastery of Twitter. It appears that the impact of social networking in the society is still evolving and will have wide-ranging implications from good to bad.

Business Model

Yet, another very important category of innovation is the business model.[32] In the early 1950s, McDonald's pioneered fast food business processes with phenomenal success. It is of interest to note that the final products contain foodstuffs known for generations: there is nothing novel or original about them. But innovations came in the way raw materials were processed and distributed, final products prepared, business transacted, customers served and payments made. McDonald's innovations have been imitated or emulated all over the world, creating millions in employment, with billions in sales and tax revenues. The fast food culture is prevalent not only in the U.S. but has also created whirlwind impacts on traditionally delicate meal services such as Chinese cuisine (see Chapter 24 of this book). There are many companies that have risen to fame due to adoption of successful business models, including Walmart, FedEx, Starbucks and others.

The knowledge economy also brings new dimensions to business models. The concept of B2B (Business to Business) and B2C (Business to Customer) are basic. Chapter 25 of this book tells the story of the Alibaba Group and its successful implementation of these models in establishing a remarkable electronic commerce empire.

Many variations and combinations are coming to the fore. One interesting model established since the new millennium is collaborative consumption, basically initiated by Airbnb.[33] It was started in February 2008 by two young schoolmates, Brian Chesky and Joe Gebbia, who moved to San Francisco and could not afford the high rent. They finally put an air mattress in the living room and turned it into a bed and breakfast operation for acquaintances. They then launched a website offering short-term living and business networking opportunities for those who were priced out of regular hotels. The business was officially launched six months later. With private investors and venture capitalists participating, it turned into a phenomenal success. The website now offers worldwide choices for simple board and breakfast for vacationing families as well as for simplified business travels. Quite unexpectedly, it provides a source of supplemental income to many families with stagnant economic prospects.

Similarly, Uber was started as a software company in 2009 by Garrett Camp and Travis Kalanick in San Francisco. It allows consumers to request car transportation or food delivery using its mobile app or mobile website. The business has proven to be a great convenience to taxi users especially at peak traffic hours or during adverse weather conditions. It also provides part-time employment opportunities for the under-employed. Thus it has become a source of economic incentive to an important sector of society.

The flourishing of business models over the internet is a unique phenomenon of the knowledge economy. The key issues to consider are the new values created and the participation of the stakeholders while the products or services are in action. Nowadays most internet users of Google maps and various websites, for example, get the service free of charge. It is the advertisers whose images keep popping up while the users are engaging the website who are paying for it. The increased focus on business models has become a new dimension of the knowledge economy and will continue to be a source of great innovation in the future.

Innovation as a Multi-talent Game

Most scientific explorations are carried out in the laboratory by scientists. Scientists like and enjoy working alone. This is the environment for incisive observations to be made and flashing inspiration to be captured. There are large-scale experiments in bio-medical, energy and engineering fields that engage a large group of researchers and technicians. The work details are generally laid out in a master game plan with connected individual objectives and time sequences. Salient scientific explorations are still to be carried out by one or a small number of people with similar backgrounds. The participating members are to observe, record and analyze the results of experimentation under coordinated or controlled environments and conditions. This is generally how inventions are initiated.

For innovation, the mode of operation is different. In the innovation process, experiments are carried out under real life conditions that are dynamic and uncontrolled. Market responses are social, psychological and unpredictable. The challenge of change is tremendous and would require the collaboration of a group of people with a variety of talents and skillsets.

Generally speaking, in addition to the initiating group who provide the core technology, it requires the support of packaging and system technologies and quality engineering. All these may be quite different from the core technology but are essential components of a commercial product. With this in place, a team consisting of marketing, sales, accounting and management expertise is needed to operate the business. Delivery of the products into the hands of the user will also require expertise in distribution, logistics, delivery and technical services. Legal and financial expertise will often be needed but this can be outsourced.

There are therefore many parties making contributions to the innovation process, not just the lead scientist or engineer. That innovation is not a one-man game is a fact often overlooked or underestimated by the scientist, the university and the government. Understanding the need for sharing and cooperation is a major challenge in research commercialization endeavors especially in the era of disruptive innovation.[34]

Chapter 3

TECHNOLOGY ENTREPRENEURSHIP

Motivation — The Spirit of Entrepreneurship

A technology entrepreneur is one who has an idea and knowledge in a particular area of science and technology and is motivated to bring it into practice for the benefit of society and mankind. The benefits may include extending human capability, creating new experience, bringing new comforts in life, solving existing health or social problems and broadening physical or social reaches.

A true entrepreneur generally has a strong sense of mission to carry out the vision for social good, not for self-enrichment. Positioning in the moral high ground is essential to sustaining the entrepreneurial effort that will be full of challenges and frustrations. Without this commitment, one would be tempted to abort the vision and retreat to an easier career path that may well bring short-term personal rewards. Thus, it should be noted that, for an entrepreneur, fame or wealth can only come at a much later stage. Personal rewards will generally come about as a by-product of achieving the ideal and vision.

The Wright Brothers

Motivation at a high level of human pursuit provides the internal strength for entrepreneurs to overcome the excruciating difficulties that they encounter. A good case in point is Wilbur and Orville Wright and the flying machine.[35]

At the turn of the 20th century, there were people in Western Europe and the U.S. pursuing the idea of flying. But none were able to stay afloat in the air for over a few seconds. The young Wright brothers were then managing a bicycle store, which was the major source of income for the family. They were attracted to the news that there were men actually exploring the possibility of flying.

Based on their experience in bicycling, they surmised that the key to stable and continuous motion was the ability to maintain balance. They observed how birds, large and small, can dance smoothly in the air at will and dart swiftly with purpose. They believed men could fly like birds via a machine they would invent that can be deftly manipulated to achieve float-ability, stability and controllability.

To confirm the basics of flying, they studied flying birds carefully and recorded the motion of the wing, head, tail and body. They made drawings and built kites and gliders with various contraptions to imitate a bird in the air.

They soon discovered that flying was very much linked with the wind. They recognized the power of the wind: how it could push, propel, stabilize the flying machine in mid-air; or hurl, twist, tumble and crash it to the ground. They experimented with flying over sandy beaches, pastures, forests, hills and valleys in varying wind conditions.

The young brothers did not have the benefit of any college education or formal training in engineering or aerodynamics, received no grants or financial subsidies from charity or government. They studied literature, searched for weather data and statistics, and sought advice from experts. They worked, toiled and occasionally put their health and life at risk.

It was hard to imagine that their initial pioneering experiments were done at Kitty Hawk, North Carolina on a strip of uninhabited wilderness near the ocean. The place was chosen after they had studied weather statistics and determined that it would be the best place in consideration of steady wind, within the reach of their limited resources, and would encounter the least amount of outside interference.

In Kitty Hawk, they fought their way through rain, storm, gusting wind, and, when the weather was calm, the onslaught of waves of mosquitos. They encountered obstacles from nature as well as government bureaucracy. But they were never discouraged enough to call it quits.

Money was an issue. It came to their mind only when they tried to support their next run of experiments. They never sought monetary rewards or thought of when to set an exit policy for maximum profit.

Luckily, they remembered to file a patent protection with the United States Patent and Trademark Office (USPTO) for their invention. They were awarded patents for the device and the mechanism of the flying machine.[36] Essentially it consisted of the three axes of motion: rolling, yawing and turning. The patents gave them some leverage in later commercial ventures but none brought them significant monetary gain. Then World War One (WWI) started and they came under government pressure to relinquish their personal interest for the higher goal of protecting the country.

The Wright Brothers were amongst the most important and outstanding entrepreneurs in human history. Their achievement was an essential element of the foundation for future air transportation and aerospace exploration. The global aircraft and airline industries of the 20th century, in significant measure, can be attributed to their inventions. They set the stage for man's endeavor to explore the moon and space beyond. Yet personal rewards for the Wright Brothers were minimal in comparison to the economic and social benefits and human inspiration they generated for generations to come.

Steve Jobs

Sixty years later, Steve Jobs was a young man with great curiosity and fascination. He had attended college for only one year. His interest was more in exploring the meaning of life than pursuing studies in the classroom. As a young man, he was fascinated with the possibility that a relatively small machine made mainly of semiconductor chips can be like a genie in the bottle for every man and woman. He had a vision that this technology would give individuals a personal tool enabling them to think and work at a level that only government and big corporations could enjoy. He wanted to make that machine to change the world.

In addition to computing power, speed and word processing capability, he made the Apple computer, and then later McIntosh, able to sit on a desktop without the need for large dedicated rooms with air conditioning. With the advent of wireless technology, the device could be made to be held in one hand, thus becoming as mobile as the man. He further added to the handheld device the ability to process music and video. Together with innovations in system integration, software development, continuous improvements and a consumer-focused business model, the iPhone that he brought to this world captivated people of all ages. The iPhone 6 today is believed to have computing power similar to the Cray 2 Supercomputer in the 1980s. So everyone can indeed have a genie in his pocket.

I have travelled many cities over many continents and invariably saw in the cabins of city transport half of the people busy looking at the smartphone in their hand, scrolling for information or talking with friends far away. It is little wonder that Apple Inc. became the world's number one in market value within 20 years of its establishment. It overtook petroleum producers, automobile manufacturers, retailing chains, banks and financial institutions and others. Nine years after the first release in June 2007, worldwide sales of the iPhone reached 1 billion units in July 2016.[167] Naturally, Steve Jobs was rewarded handsomely through the stocks he held as founder of the corporation. But it is fair to say that money or wealth was not what he had set out to seek.

Steve Jobs wanted to change the world. With this spirit, he challenged and indeed pushed his colleagues to strive for the impossible to make a product "perfect" for the people. He was criticized for his "distortion field" mentality and harsh attitude towards some colleagues. But, basically, Jobs set very high and sometimes unreasonable standards for his colleagues as for himself.[13] In persuading John Scully who hesitated to quit his presidency at PepsiCo to join Apple, Jobs was quoted as saying, "Do you want to continue making that sugar water or join me to change the world?" Scully got on board. The Mac computer, iPads, iPhones and most of what the company made under Jobs' leadership have changed the lifestyle and work style of mankind for generations to come. The world!

Ma Yun

Another 20 years later, Ma Yun, known as Jack Ma outside China, is an entrepreneur of uncommon accomplishment. As a young boy from a modest family in Hangzhou, he had a fascination with the English language. He studied it diligently on his own and sought to mingle with visitors who came to the scenic West Lake so as to get opportunities to learn about the outside world, to practice conversation and improve his proficiency in English. His diligence and sincerity enabled him to establish congenial friendships with many foreigners. One of them invited him to visit Australia around 1985. The trip proved to be an eye opener for the young Jack.

He had been under the impression that life in socialist China was like paradise and that, under the capitalist system, the Western world was in grave danger for its very survival. He realized that it might have been China that was in peril in competition with the Western world.

He developed a strong sense of responsibility to help modernize and revive the Chinese economy.[37] He decided to focus on small and medium-sized enterprises (SMEs), which are the major job providers in Chinese society. The SMEs were a link to the livelihood of the people and should therefore get acquainted with the West to expand opportunities. He sought diligently to discover how.

In 1995, he had an opportunity to visit the U.S. and came across the internet for the first time. He was awed at its power and thought this would give the SMEs a great tool to connect with the outside world and advance their business. On returning to China he tried a number of approaches. Finally in 1999, he established Alibaba Online with 17 friends. He used Alibaba.com as a platform to provide B2B services. This would enable or facilitate individual producers in China to connect with buyers in the West to conduct business.

After two difficult years of starting up, the SME producers gradually gained familiarity with and trust in the operation of the B2B platform. He then established Taobao.com as a C2C (Customer to Customer) electronic commerce platform.

In 2004, Alipay was established as a third-party online payment platform with no transaction fees and proved to be phenomenally successful. Through these Alibaba platforms, Ma established a value chain to support the e-business system and has benefitted millions of SMEs in China.

All 18 co-founders of Alibaba.com became multi-millionaires but wealth came at the end of an uncertain journey. It was the motivation to help the SMEs and the society that carried them forward (see Chapter 25 of this book).

To Be or Not To Be

In closing, I would like to mention that I have often been asked by students how one can make one's first bucket of gold by going into entrepreneurship. I replied that if someone is really interested in making money for himself alone, he would be better off doing something else: taking a stable job in business, playing at card tables, betting on horses, trading stocks or buying and managing real properties, etc. Pursuing technology entrepreneurship to get rich is a risky and tortuous process. The odds of success are relatively low. To pursue entrepreneurship, one needs to set higher goals in life and take personal rewards off the mind. This will enable one to endure the pains that come along the way. That is the essence of the entrepreneurial spirit.

Chapter 4

BUSINESS MODEL AND BUSINESS PLAN

Business Model: Why and What

What is a business model? Why does one need to consider business models? Why is a novel technology or unique product not enough for business development? These are questions frequently asked by budding entrepreneurs and accomplished scientists alike.

"Business model" relates to how one delivers the product or service to customers, create the value as promised, make them promptly pay as agreed, and continue with further business. "Business model" directly relates to the revenue stream and the sustainability of a venture regardless of the uniqueness or novelty of the product.

In my consultations with accomplished scientists thinking of going into entrepreneurship I usually asked bluntly about their business models. What? Why? To my surprised colleagues, I usually refer to an example in food technology — the making and selling of dumplings.

Yes, you have the best technologies and know-how for preparing all the ingredients and the processing techniques to make the world's most amazing dumpling. But at the end of the day, how are you going to sell them and make your customers pay you for what they have received? This involves additional business preparation and investments that are often larger than that for the core technology.

Now, if you want to sell your dumplings by opening a dumpling shop, you need to find a very good location for the shop, fit it out and find someone to run it. You will need to train the person(s) very well and make sure they can work long hours. The customers will pay on site for sure but the income will be relatively small unless you can multiply the number of stores very quickly. Alternatively, you can sell by cooking the dumplings and filling them in cans or refrigerating them and placing them in vacuum bags. Then you will need investments in canning, refrigeration, packaging, shipping and delivery, and most importantly, arrangement with grocery stores for display shelf space. These stores are known to have long payment cycles and you will need to have enough cash reserve to sustain the operation. In the end, you will find that your core technology was only a small portion of the entire business value chain.

So what is your business model and business plan? The concept of business model was practiced since the early 20th century. The "bait and hook business model" was a classic for many industrial products. This involves offering a basic product at very low price and charging for refills or replenishments for long periods, which was really the revenue stream. Examples are razors and blades, computer printers and ink cartridges, instant cameras and photographic film, etc. However, in the digital age the business model has become more complex. It involves the value structure and its delivery in the business value chain.[38]

The modes of operation in latter day electronic commerce, B2B, B2C or C2C are useful tools in analyzing business models.

In the industrial economy, the producer's enterprise sells manufactured products to the consumers. This is a simple business model of "Business to Customer", or B2C. The consumer pays the producer or its sales agent and completes the deal. While examples are numerous, a paint company selling house paint to homemakers to paint their houses is an obvious one. The most complicated segments in this model are the distribution channel and post-sales service.

Alternatively, the producer can sell the products to another business which will use it as an input in its own products, or B2B. An example is when the same paint company sells a paint formula to an auto manufacturer to paint its cars to sell to consumers. Obviously, the product will be quite complex since the final application processes and performance requirements are very different. The recipient business will pay the producer according to the terms of the sales contract.

In these two examples the business entity producing paints will have to consider the product formulations even if they are of the same generic type.

This includes technology development, methods and conditions of application, properties as required in performance, selling mechanism, delivery scheme, logistics management, advertisement and marketing costs, post-sales service and payment schemes, etc. All these will affect the profitability and the sustainability of the company.

Some manufacturing companies may develop specialties in the B2B business model while others may have expertise in B2C. In the worldwide chemical industry, for example, the Du Pont Company has been known to be strong in B2B; while Colgate-Palmolive has developed expertise in B2C. Some more recent examples will be discussed later.

Microelectronics — TSMC

It is essential for a manufacturing company to study the nature of its business and customer needs and how it can create value for the customer. Then a business model should be constructed and strategy formulated to develop competitive advantage using the business model. The Taiwan Semiconductor Manufacturing Corp (TSMC) is a case in point.

TSMC manufactures microelectronic circuits with various performance properties for integrated device manufacturers (IDMs), in a typical B2B business model. But, instead of making the same chip for every user, it studied the specific needs of each customer, for instance, Texas Instruments Incorporated (TI).

An established giant in the semiconductor business, TI has an excellent technology base but its manufacturing was costly because of local issues. TSMC proposed to TI that they would make TI's chips with identical performance and quality to TI's own manufacturing division. TSMC will be committed to a specified delivery schedule, and at a lower price. Further, they would not sell the same chips to any other computer manufacturer nor would they make their own brand of devices to compete with TI's. Obviously, to TI executives who had been concerned with problems in labor and cost from its own manufacturing division, this was music to the ears. After a trial period, TI was happy to contract TSMC for the IC it needed.

In similar fashion, TSMC signed up IBM, Hewlett-Packard, Apple and other computer makers as their supplier. Within 10 years of its establishment, TSMC became the world's number one supplier of semiconductor chips and premier foundry of semiconductors. Its business revenue grew steadily despite several global financial crises. After the year 2000, it expanded its business into servers, smartphones, wireless devices and cloud computing infrastructures and applications.

TSMC started as a spin-off company of the Industrial Technology Research Institute (ITRI) in Taiwan in 1987 and was supported by ITRI trained engineers and professionals (see Chapter 17 of this book). However, its business model and business strategy are important innovations and key to its success. These were attributed to the leadership of Morris Chang,[39] founding Chief Executive Officer (CEO), one of the brightest brains in business strategy in the semiconductor industry. Before coming to ITRI, Chang had significant executive experience in the U.S. including at TI and later, General Instruments. In the mid-1980s, when Taiwan was striving to transform to a technology intensive economy, he was recruited to ITRI as president and, three years later, decided to lead the TSMC entrepreneurial venture. At that time most investors were hesitant as to the future of IC and the NT$6 billion target to start the company (approximately US$200 million) was staggering. Fortunately, the government under Chiang Ching-kuo (蔣經國) was visionary and stood strongly behind it. The National Development Fund injected funds to take 25% of the equity. But this was still far off the mark. Chang took it as his responsibility to persuade other investors including Philips Electronics of the Netherlands, who became the major investing partner.

To support its competitive advantage in chip manufacturing, which could rapidly change as advanced materials technology and manufacturing processes progressed, TSMC invested significant resources in research and development. It focused on nanoelectronics and related manufacturing technologies. Once its technical target was attained, the company followed with capital investment to upgrade its manufacturing capability.

From 2009 to 2016, TSMC's research and development (R&D) increased by 220% to reach US$2.2 billion while capital investment increased by 250% to reach US$9.5 billion. But its unit price on 12-inch wafers only increased by about 5%.[40] Thus, over this period TSMC overtook Samsung and other competitors to excel in the fabrication of 28 nanometer (nm) and 16 nm Integrated Circuits (ICs). For 28 nm wafer, TSMC has a world market occupancy of 80%. It is now vying for the number one position with Intel in 7 nm technology. By the third quarter of 2016 its market value reached US$160 billion, at the same level as Intel's.

To sum up, under Chang's leadership the company operates on a business model that has several dimensions:

- Strategy: on a business to business framework create specific value for specific clients in quality and service.
- Focus: stay in area where it has competitive strength; never wander from its core capability.
- Nurture excellence: Invest in R&D in the fields chosen to maintain leadership in leading edge technology.
- Execution: upgrade fab manufacturing capability once R&D results demonstrate feasibility.
- Respect and reward: treat employees with stock options and bonus to share company profit.

Electronic Commerce — Alibaba.com

Creating value for the customer is the crux of a business model. Generally speaking, technology entrepreneurs with a strong background in science and technology are prone to focus on the novelty, quality and uniqueness of the technology the company provides. A non-technical person, on the other hand, is more likely to view the business from the perspectives of value chain and customer satisfaction. A good example is Jack Ma and his design of the Alibaba Group, which put into use several business models.

- Business to business, B2B
  In 1999, Jack Ma established Alibaba.com to promote B2B commerce between product makers in China and buyers all over the world. It was Ma's vision that business would rapidly flourish once the two sides met. He was quick to zero in on the window of opportunity.

- Customer to customer, C2C
  In 2003, it introduced Taobao.com for C2C business. This was after customer confidence with Alibaba B2B and with China joining the WTO. Like with the B2B platform, there was no charge for users but incomes from advertisement and related services surged to make it a success.

- Business to customer, B2C
  Within two years, Alibaba set up TMall.com as a B2C website for brand name products to reach the consumers in China who were rapidly growing more affluent.

These are the basic business models set up to capture the vast e-commerce markets. A more detailed discussion of Alibaba appears in Chapter 25 of this book.

Business Plan

The purpose of the business plan is to state the business purpose, long- and short-term business goals, the basis of the goals and actions to be taken to achieve these goals. It should be done after the business model is decided. Various components of the business should be outlined in the business plan: market needs, targeted customers, products or services, business strategy, source of technologies, manufacturing capacity, sales channels, environmental constraints, financial analysis and forecasts, human resources requirements, key management team, key mile-posts, anticipated risks and the criteria for measuring success. These elements will define the quality of the business plan.[41]

Thus the business plan shows the interrelationship of various elements of the business, how these elements combine to realize the goal of the company and produce profit to sustain it. The business plan should be understood and agreed to by all the members of the venture to guarantee their commitment.

Working to finalize the business plan will provide the team members with opportunities to analyze and discuss the business venture from many different perspectives. The issues discussed will most likely include the following:

- What niched advantage will the product/service bring to the customer — novelty, performance, quality, price, delivery, etc.?
- Can these be demonstrated?
  Quantify the advantage to the extent possible or set clear benchmarks to quantify.
- What kind of users will most appreciate the advantage or value?
  Different users will perceive the value differently; understanding this will build a strong customer base and revenue stream.

- What is the size of the market — age, gender, geography, special interest, etc.?
- How much will the users be willing to pay for the value-in-use?
  Do we have freedom to set price?
  Who are in the value chain to share the revenue?
- How can the value be best delivered to the user — in what form or package? In what time or sequence?
- How will the user pay for the value?
- What is the end game; review the business model.
- Who are our competitors — current and potential?
- What are the possible roadblocks and risks — law, capital, people, intellectual property rights?
- Are there gaps in our technology?
  If not the core technology, what about packaging, quality or system technology?
- What will be our monthly burn-rate?
  This should include salaries, rent, utilities, consumables, travel and administrative support.
- How long will it take to demonstrate the uniqueness of our product?
  List the work breakdown structure. Which ones are key determining steps?
- When and who should we go to for the next round of fundraising?
  List potential investors and how to approach them.
- Who will run the company full-time?
  Part-time professors/advisors are not good enough to run the company. The company needs a leader round the clock.
- Are there gaps in the skillsets held by the team members?
  Who is going to scout the market and find buyers, negotiate sales contracts and collect payments?

Going through the list may be a humbling and sobering experience. Obviously there will be more issues raised as the discussion proceeds. And, generally speaking, there will not be full answers, but the aim should be to construct the best available business plan for the time being.

An added benefit of the business plan is that it will enhance cohesion of the team. After the plan is finalized, the mutual dependence of team members will become apparent. Each member of the team should be able to articulate the overall business plan.

The business plan is a working document and subject to review and revision at proper periods of time. As such, it should be dated as it reflects the status of the company at that time.

The business plan can take many forms with different levels of detail depending on the audience it intends to address. If the audience is the investor, it should use plain language to present clearly the challenges and choices for their understanding and commitment. It is better for the investor to know the possible risks and liabilities so there will be no surprises in the future.

A business plan will generally include a vision on the possible exit mechanisms: when and how can the investors recoup their investment? Investors are interested in knowing their actual risks and when they will have the freedom to move on to other opportunities.

The business plan can be as complete as is needed and is a confidential document. Several versions can be prepared. The external version, which can be used for investors should be no more than five pages with an executive summary. Detailed and confidential data may be attached as appendices as appropriate.

Chapter 5

THE INNOVATION PROCESS

From Lab to Market

In the course of the last century many industrial corporations in developed countries set up research laboratories to identify unique research opportunities for commercialization. Corporate giants such as General Electric, IBM, E.I. du Pont de Nemours, Eastman Kodak, Fujitsu, Hitachi, Siemens, Ericsson, Merck Sharp and Dohme, and the like, have actively pursued research in their fields of interest. In parallel with this, research universities such as the ivy leagues and the leading state universities of California, Michigan, Illinois, Wisconsin, Minnesota, Virginia, Maryland, New York, etc., working by themselves or in collaboration with industrial partners, are also active in scientific research. Thus management of research, technology and innovation has been an important subject for business and academic institutions alike. Management science is important because it deals with knowledge workers who are key to the competitiveness of organizations. There is a good deal of literature on the subject. A comprehensive discussion with abundant references can be found in Frederick Betz' book[42] and many business review publications.

How is the result of scientific research successfully commercialized to produce value for society? To illustrate the salient features of the innovation process, we will take a basic case of scientific research from the university. From this, the relationship between university (U), technology institute (T), government (G) and business (B), as well as the challenges and choices of entrepreneurs in the wealth creation process will become evident.

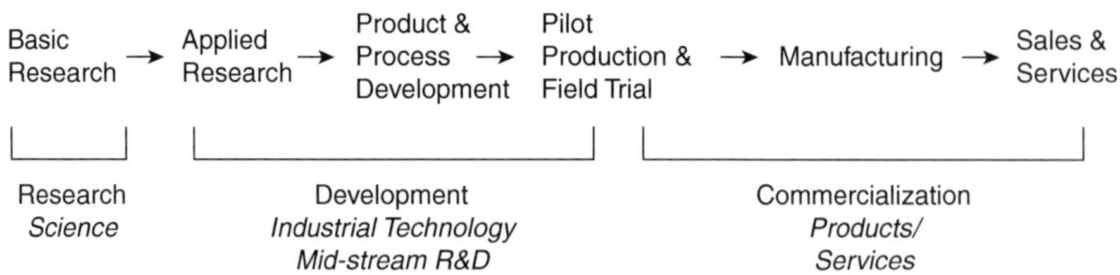

Figure 4. From Research to Development to Commercialization

The process of research to commercialization generally starts from basic research to applied research, product development, process development, pilot production, pilot market trial, manufacturing, distribution, sales and post-sale services (see Figure 4).

The process, normally, can run in any direction: bottom-up, from left to right, referred to as technology push; or top-down, from right to left, referred to as market pull. Each process has its pros and cons.

In technology push, the process of commercialization is usually driven by the technology leader. Every step along the way can more or less be forecast and is thus easy to manage. The pitfall is that the product finally developed may not match real market demand: it may be too early or too late. A product may be too early due to overenthusiasm for the technology. The market has not yet caught fire. So you may be selling a solution before the problem is fully identified. A product may be too late generally due to competition. Your competitor has found a way to be in the market before you.

In market pull, the process is generally led by the marketing group who is close to the market but may not have an adequate understanding of the development process and the technology needed. As it moves along, some gaps in the process may appear. Thus it will take more time and effort to develop or find other sources to close the gap. This may make the product too late or too costly for the market.

Both these approaches are standard R&D practices in an existing company. It is generally the higher levels of management who decide which approach to take. In reality, the research commercialization process can start at any point, return to basics or to an earlier step and spiral to the end.[42] In some cases it can start at several points in parallel. Depending on the technology and market, some steps can be shortened considerably, especially with the advent of information technology.

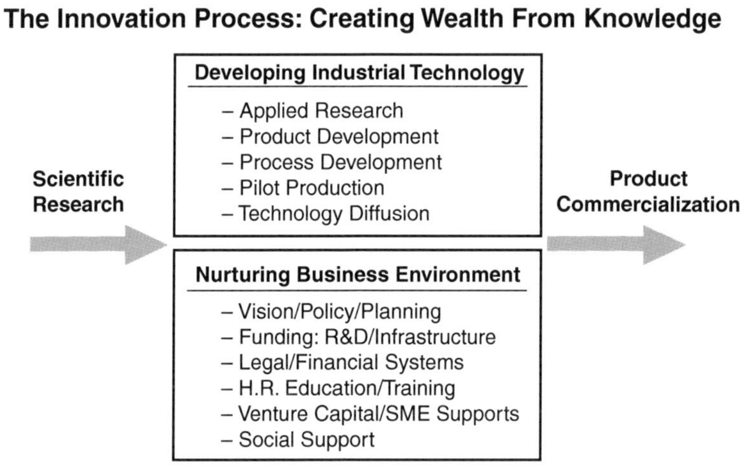

Figure 5. Innovation: Wealth Creation Process

From a systems point of view, the research commercialization process consists of two interlinking components: development of industrial technology and nurturing of business development. This is shown in Figure 5.

It should be clear that the innovation process is a complicated process that will require the cooperation of many players: both groups and individuals. The relationship of these players is shown in Figure 6.

Note that technology institute (T) and government (G) are the bridges linking university (U) to business (B).

The university (U) generally undertakes basic research with the important objective of advancing human understanding of the universe, its environment, inhabitants and interactions, thus creating new knowledge. However, not all basic research has immediate application. To trigger the process of commercialization, the research must display prospects of creating value in certain end-use applications in the foreseeable future.

If the research is concerned with the synthesis of a particular material or device, it must be shown that the material or device can be synthesized repeatedly with reasonable quality and quantity. It should be noted that in basic research the question of "yield" is not a major concern. So long as the result can be demonstrated and confirmed, the researcher can go ahead and publish results in scientific journals and be credited with the finding. This satisfies the objective of the university in the creation of knowledge

Players of the National Innovation System

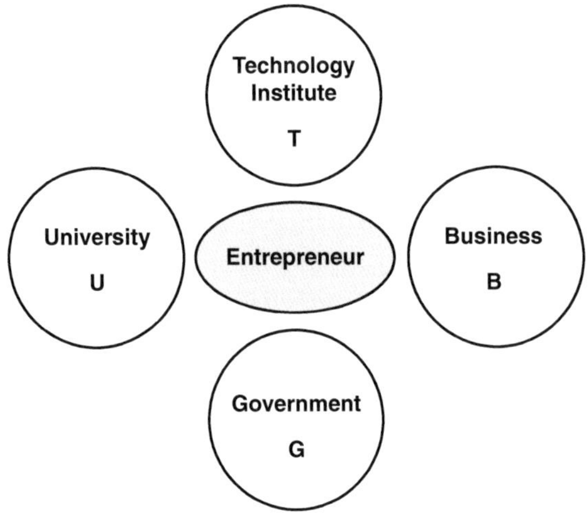

Figure 6. The Players of the Innovation System

through research. But for commercialization the yield is a key factor of cost and profitability, which is the issue of survival for a business.

The development stage is usually repetitive and costly and offers relatively little scientific interest to the researcher. But it is necessary for the process of commercialization. Therefore development work is a group effort and best suited to a technology institute (T) whose expertise is to develop the necessary industrial technology with satisfactory efficiency. The institute should and can dedicate full time to work on this. Often, the process of development is a source of further innovations, which generally result in patents or know-how that are the intellectual property of the developer.

While patenting gives the inventor some legal protection and exclusive rights in commercialization, it nevertheless spreads new knowledge in the public. Some companies may choose to keep this knowledge to itself as know-how, and not through patents. This is a risk to be decided by the company. The U or T would sometimes apply for a patent first, or make adequate preparation for patenting, before going forward to publication. This is because recognition as "first to discover" signifies leadership and has its merits.

From Core Technology to Value-in-Use

Another way to look at the wealth creation process is from the perspective of value. For a technology to be of practical use, the core technology, material or device must be embedded in a form or package deliverable to users for application. To accomplish this, some system of integration or packaging technology will be needed. This is the stage of product and process development, generally the realm of industry technology.

A recent example is the invention of the chemical compound graphene, whose novelty and uniqueness merited the award of a Nobel Prize in Physics in 2010.[43] It has shown great promise in energy conversion, energy storage and engineering applications that require ultra-mechanical strength. But the technology related to its application remained to be developed.

There were many similar earlier Nobel hopefuls such as Buckminster fullerene C60, whose inventor was awarded the Nobel Prize in Chemistry in 1996[44] and High Tc Superconductor, the Nobel Prize in Physics in 1987.[45] The great promise they showed and the frenzy each created in microelectronics, energy and environmental applications still remain a promise today. It is fair to say that whereas the scientific significance in these cases were quickly recognized, the challenge of innovation or developing industrial technology was, in general, greatly underestimated.

The development process differs in complexity with the nature of the product. For biochemical, chemical, electrical, energy, material and mechanical fields, the process can be lengthy and protracted. The support needed are socio-economical, far more than financial.

A company needs to be established (B) with team members in place and the principal ready to champion the project. Possible business models should be explored and a business plan established. The plan will pull together all facets of business management: marketing, financing, sales and distribution, logistics and supply chains, and after sales services. It is only then that the product can be delivered to the final users who will illustrate the value-in-use of the technology. This process is shown in Figure 7.

This graph shows that the core technology needs to be manufactured, incorporated into a system and properly packaged, with quality assurance before it can be called a realistic product. The product needs to be tested in the market, find its distribution channel, make sales and, if the financials are in order, then it has a business. It is after all these steps are taken that the core technology can demonstrate its end-use value to the customer. It is a long process far beyond the capability and endurance of the inventor scientist and engineer alone. This is the reason why most entrepreneurial efforts have failed prematurely.

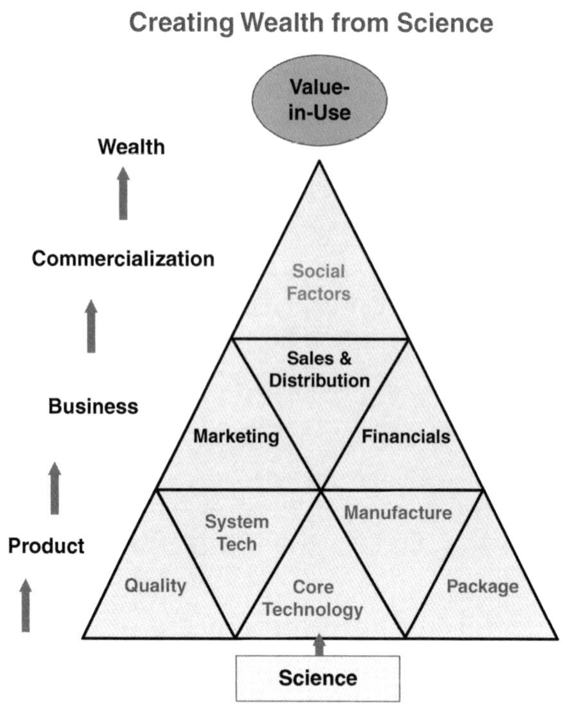

Figure 7. From Core Technology to Value-in-Use

The resources of the technology institute can be called on to provide assistance. This will most probably lead to the participation of venture capitalists, finance managers, logistics and supply chain partners or business collaborators. Taking this integrated approach with the support of the government is the choice of many newly developing economies.

For software projects where hardware requirements are at minimum this process can be considerably shortened and simplified. This makes software-oriented ventures easier to commercialize in the knowledge economy. In the last two decades, various software companies have sprung up on university campuses and technology parks. These are popularly called incubators.

Incubators require relatively small spaces and some seed funds. They will facilitate the inventors to go for a try. However, for software ventures, the efforts in debugging and in establishing compatibility, scalability, upgradability, user friendliness, etc., can be tedious. Expanded scale of market trials with possible modifications of the products and/or services will thus take place.

Young start-up companies generally need support in terms of financial, market, network and management resources. Therefore the ability to provide good mentors is an important factor for a successful incubator. On the Chinese mainland alone, it was estimated that over 4,500 incubators were operating around 2010.[46] Larger incubators that can provide more space and more resources are generally called accelerators. They are generally run by the government or major corporations such as Microsoft.[47]

Again, this commercialization process can be long and drawn out depending on the nature of the project and the business environment. If the institute (T) is behind it, more contacts can be explored and interested parties in the existing business community (B) can be enlisted for support, either as a technology recipient or investment partner. Normally, the business model and business plan originally established are adjusted and modified as the process moves along. If the eco-system is properly constituted, there will be ample expertise in the technology and business community to provide assistance to the entrepreneur.

Bringing new drugs or pharmaceuticals to the market involves a different set of problems. This requires large scale clinical tests for compliance with government regulations and presents many difficult challenges for new start-ups because of the expertise and expenses required. A common trend is for the start-up to partner with established drug companies which already have the capital, channels, experience and resources needed. Normally, when the efficacy of the new compound is scientifically demonstrated, venture capitalists will come in and big name drug companies will be persuaded to join. This is a field in which university, institute, government and many types of business corporations can participate.

In the last two decades the leadership of Massachusetts Institute of Technology (MIT), Harvard University and Massachusetts General Hospital have nurtured the Boston area into becoming a center of bioscience in northeast U.S. Similarly, University of California-Berkeley (UC-Berkeley), University of California-San Francisco (UC-San Francisco) and Stanford University have brought new vitality to the San Francisco Bay area with bio-medical research and innovations.

It is essential that high standards of personal conduct and business ethics should be observed by all parties in the process. A 2016 case in Taipei has implicated the President of the Academy of Science who was finally pressured to resign on account of transgressing the line of professional conduct and business ethics.[48]

For the start-up business to operate successfully, all management functions have to be taken into consideration, including product manufacturing, R&D, product strategy and planning, financial management and human resources development. This could very well be outside the skillset or temperament of the scientist whose pioneering work kicked-started the project. Thus the lead scientist/engineer is not necessarily the best choice as team champion. The champion should be one with entrepreneurial spirit who can work among the institutions and various personalities to accomplish the goal. The unfortunate saga of Shockley Semiconductor Laboratory of Silicon Valley in the mid-1950s, involving a Nobel Laureate and the inventor of the solid state transistor, was a sobering lesson.[49] Interested readers are referred to Chapter 16 of this book.

Therefore, the research commercialization process is not for one person, but requires the concerted efforts of multi-skilled talents. It is not surprising that Steve Jobs needed the support of Steve Wozniak and John Scurry, as did Bill Gates with Paul Allen and Steve Palmer.[50]

In assisting the research commercialization process, the four institutional players, U, T, G and B, should complement each other. They have expertise in different fields. This is a feature one will find in a country or city conducive to the development of innovation and entrepreneurship. Some examples will be given in Part Four of this book.

Part Two
Making It Work

Inside a city, region or country, one will be interested in learning:

- *How do the institutional players interact with each other?*
- *Do they see each other as colleagues or as rivals?*
- *Do they work to complement each other?*
- *Do they join hands to support individual entrepreneurs?*

The mode of operation of the institutional players is the central issue that determines the effectiveness of an innovation system, be it national or regional.

Chapter 6

THE NATIONAL INNOVATION SYSTEM

The Four Institutional Players

From the middle 20th century, technology communities in developed countries have paid great attention to expediting the journey from research to commercialization. Innovation efforts in the U.S. and Western Europe were led mainly by industry. Major corporations such as General Electric, IBM, Exxon, AT&T, Du Pont, General Motors, Boeing, Xerox, SmithKline (now GlaxoSmithKline), Ericsson, Philips and Siemens established R&D divisions to pursue their commercial interests.

Since knowledge creation has wide impacts on social and human behaviors, the academic community has also shown great interest in the subject of the national innovation system. Research work was pioneered by R.R. Nelson at Columbia University,[51] B-A Lundvall of Aalborg University[52] and C. Freeman of the University of Sussex.[53] In an effort to promote innovation as a driver of the new economy, the Organization for Economic Cooperation and Development (OECD) has devoted special effort to coordinating and assisting its member states in innovation.[54,55] A recent report outlining the development process of innovation systems inside the OECD system was given by Godin.[150]

To the surprise of the world, in the late 1980–1990s, several East Asian economies did remarkably well in building economic power through developing technology and innovation. There was growing interest in finding out how this was accomplished since, in the region, both industry and universities have been relatively weak. Henry S. Rowen of Stanford University conducted a systematic study involving many researchers on the social and political foundations of the East Asian growth. The role of government and institutes have received special attention.[56]

In 2007, the Salzburg Seminar (later called Salzburg Global Seminar) had undertaken the initiative to explore the "Lab to Market" process and the role of partnership in accelerating innovation.[57] This was followed in 2008 by another session with wide European, U.S. and East Asian participation to illustrate the relationship of innovation with the knowledge-based economy.[58]

We will now take an in-depth look at the partnership of university (U), business (B), government (G) and technical institute (T).

The process U to B is frequently pursued by universities. The results are often disappointing. This should not be a surprise since university and business are two types of organizations with different objectives, skillsets and criteria for measurement of success. A successful direct transfer of university research to commercialization is really more the exception than the norm.

Many governments learned this the hard way. Accordingly, some leaders would feel that the government has to take a direct leadership role in the technology commercialization process for G to B.

But this will not work for the same reason U to B will not work. Besides, direct involvement of the government with business may result in conflicts of interest. Hence the process G, U to B was not found to be any better.

As an alternative approach, some governments tried to set up technology institutes to drive the commercialization process: G, T to B. The three party framework has been referred as the Tri-Helix model.[59] It is an important concept but there have been few major success stories to brag about.

An analysis of the research commercialization process presented in Chapter 5 clearly showed that the four institutional players, U, T, G and B, all have a unique role to play. Thus, the process cannot be fulfilled by one player, or two or three players. Close interaction of all four is required to provide a firm framework and eco-system for entrepreneurs to strive and the knowledge economy to prosper. The four are different in nature, mission, function and performance standards. But each can bring in unique ingredients. They are the four pillars of the national innovation system. Thus we should pay attention to the ways these institutions interact.

To facilitate understanding of the subject, we can distinguish three modes of interaction among the players.

Separate and Independent

The first mode is characterized as separate and independent. Here, each player works towards its own objectives, undertakes programs at its own pace and has its own criteria for the measurement of success.

The university, for example, naturally focuses on attaining excellence in teaching and research. It seeks funding support from the government and industry to pursue its research interests and furnishes reports to the funding organization in time as the contract requires. But the university generally does not understand or is not able to address the seriousness or urgency of the problems that industry encounters. As a consequence, industry finds university research of little or no value in solving its pressing problems. The industry would, also quite naturally, view university people as living in an ivory tower and insensitive to the real world. The students may be accomplished and creative but the industry may find them lacking in the aptitude and skillsets that are needed for the real job.

The government, on the other hand, provides funds to institutions of higher learning to educate its citizens as an obligation. It collects taxes and duties from industry and business to fund its many obligations. Other than this, the government tends to stay away from industry, leaving whatever problems industry has encountered to its own solution. The government believes that market forces will direct industry to settle its problems in due course.

Industry would also be apprehensive if the government becomes overly active in extending its role for fear that it may be interfering with their normal business operations. It is the prevalent feeling that, if left to bureaucrats, things would get worse rather than better, even much worse. The university, on the other hand, is generally sensitive about upholding academic freedom. It would rather see the government at no time other than funding time. It is therefore no surprise that the government tends to adopt a *laissez faire* policy.

Technology institutes, in these circumstances, have very limited functions. On the one hand, some institutes may be academies of high standing and their roles only of honorary or advisory nature. On the other hand, institutes may only be good enough to play specific roles such as in accreditation, certification, standardization, quality assurance and technological liaison. It is inconceivable that technology institutes would serve other useful purposes.

Thus, all parties mind their own business and play by their own rules. This is the natural mode of interaction for institutions that can generally be found in a developed market economy. One may say that separate and independent is the default mode of interaction for institutional players.

Eighteenth century Western Europe provides a good example of this model. Each university was proud of its own mission and tradition. The technology institute did not exist or existed only within its own confines. Businesses had freedom to exploit the situation fully and the government sat on the sidelines. In this model, innovation exists but is not active and is mostly run by established businesses. As an example, the concept of the steam turbine was known in the early 17th century. But it was not until 1781 that James Watt patented a steam engine with stable rotary motion suitable for driving factory machinery with about 10 horsepower. It would take another 50 years for further improvements and development of new applications in European countries to bring about the industrial revolution.

Fully Integrated

The opposite of the above is a system that is fully integrated. This is common for countries operating under planned economy doctrines. Here, the government makes the overall development plan and funds the universities that may also operate the technology institute. In fact, the government may establish certain mission-oriented institutes to accomplish specific national goals in business. When a research project is completed, some scientists or engineers take over the implementation and commercialization. It is an operating system designed for efficiency.

But there are problems with this mode of operation. First, the project starts with a top-down goal and timeline. This would leave very limited room for flexibility and ingenuity, which are essential for new value to be created. Second, completion of the technical part will not guarantee commercial success because market acceptance of the completed product or service is generally only an after-thought. Third, there are questions on the nature of operation that may hinder the commercialization process: conflicts of interest, overlapping roles of player and referee, confused owner and manager relationship, etc.

In the post World War Two (WWII) period, Soviet Russia demonstrated many extraordinary technical achievements. In October 1957 it was the first nation to launch a satellite into the earth's orbit.[60] Capturing the fancy of the world, the 1,500 kg satellite, Sputnik 1, managed to spend three months orbiting the earth. It triggered many new scientific, technological and military developments. This was a Russian government effort involving many ministries and state industries. The Russian Academy of Sciences was believed to be playing the scientific leadership role. Obviously this historical launch gave Russia an early lead in the space race.

However, successful completion of the Sputnik Program had not given Russia any advantage in economic development. The economy remained depressed which, among other factors, led to the fall of the Soviet Union 30 years later. In a nutshell, the national innovation system of Soviet Russia was skewed towards the government with little consideration of business and economy. Although it was successful in accomplishing certain national objectives, it failed to create new economic value for the people.

Similar observations can be made in a few other countries. A case in point is the reported development of nuclear technology in North Korea. By the end of 2016 six nuclear tests had been carried out by the government. On July 7, 2017, its Supreme Leader Kim Jong-un announced the successful launching of a midranged Inter-Continental Ballistic Missile (ICBM) that travelled nearly 1,000 kilometers (km) down range and landed in the North Japan Sea. Experts have estimated that North Korea has developed capability in both thermo-nuclear bomb and delivery system that can post threats to continental United States. This has resulted in a fiery shouting match between Kim and U.S. President Donald Trump while the rest of the world was hoping to see a calm and constrained atmosphere in resolving long-standing international problems. [179] In any case, it appeared that North Korea must have assembled a substantial group of its expertise through its university and technology institutes managed by the government.[61] However, the economy of North Korea remained depressed and, with a GDP per capita of under US$1,000, the majority of the population existed under the poverty line. What would be better for the people of North Korea: decent livelihood with steady growth or temporal glamour with the risk of war? This is the question for the North Korean people to decide.

In summary, the fully integrated national innovation system is one where the university, institute and business surrender to government initiatives. They become policy tools of the government. The national innovation system can be effective and powerful in attaining government objectives, but is not one that is conducive to innovation and entrepreneurship in the commercial world.

Positioned and Coordinated

This is a system where the four institutional players: government, university, institute and business, have defined missions and positions. They respect each other's position and role. There may be minor overlaps in their program objectives, but these are not duplicates and will serve as linkages to reach out towards each other.[62]

In an innovation system that fits this mode, the university focuses on conducting basic research and nurturing students with inquisitiveness and creativity. The technology institute converts the research results to technology that can be measured for product quality and product yield. It can also work with industry/business full-time to address urgent issues. The finished technology can be transferred to the existing business as an addition to their business lines. Alternatively it can be transferred to a new group of entrepreneurs or become a spin-off of the institute or university to pursue commercialization. It should be noted that the resulting commercial product could be a product or service that is not exactly as originally conceived, but the core technology will have been established and the entrepreneurial team can make adjustments to it as market demand changes.

In this system, the government provides support to the infrastructure, which includes both hardware and software. Hardware infrastructure includes factory and laboratory facilities; software infrastructure includes human resources, legal system, venture capital, financial and management services. Each institutional player performs its part as in a quartet concerto playing good music. This is a system that performs with the highest level of efficiency.

With such a system in operation, an individual entrepreneur can find resources and support from any or all of the institutional players. The entrepreneur then becomes the common thread that links the parts and activates the system. With good music playing, the dancer will dance to his/her potential. This will represent a functioning innovation system.[63] An illustration is shown in Figure 8.

The System Factors of Innovation

The government should take the lead in establishing a national innovation system. It should also take action to nurture the environment so that the system is conducive to innovation and the growth of technology-based

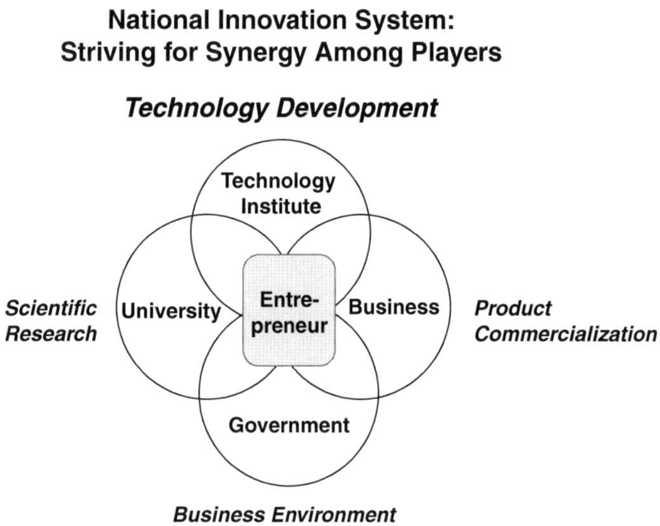

Figure 8. A Functioning National Innovation System

businesses. An effective national innovation system is one that has institutional players in place with well-defined roles and a clear mode of operation.

It should be noted that the word "national" in national innovation system does not refer to "nation" in the political sense. It can refer to any city, province or region, in whole or in parts, whose economic development is knitted together in such a way that can be considered as a self-sustaining entity. The term is used in the sense of "macro-economy," which refers to the whole, in contrast to "micro-economy," which refers to the individual or the firm.

The system factors, or environmental factors of the eco-system, are issues of common interest to the players whose resolution will facilitate their collaboration. The process of seeking consensus and resolution will foster the spirit of innovation. These issues can be analyzed via several categories, as follows.

- **Vision and Strategy: Planning, forecast, policy and program**

  A vision of the city or nation, and the strategy for achieving it, should be formed and articulated. This will involve a systematic and periodic review of Strength-Weakness-Opportunity-Threat, followed by updated forecasts, reviews of policies, allocation of resources and establishment of new policy and programs. In the last 30 years, the development of some Asian economies clearly followed this approach.[64,80,129] It should be noted that for this to function properly, all the institutional players should be involved in the process of arriving at consensus and ownership.

- **Developing Capacity: Human Resources (HR), infrastructure, technology and processes**

  A fundamental function of the innovation system is developing technology and innovation. It thus requires the assessment of key capacity: capability of the institutions, basic infrastructure needs, human resource needs at different levels, technology forecasts, etc. The university will undoubtedly play a pivotal role here.[65,66]

- **Technology Diffusion: Upgrading existing industry**

  There is a need to review and establish legal guidance for technology diffusion. This will facilitate technology business formation. It should reward researchers and developers, motivate investors and provide justice to the general public.

- **Expediting Research Commercialization and Incubation: Clusters, network, Intellectual Property Rights (IPR), venture capital**

  This is to foster an infrastructure conducive to the growth of technology business including: technology marketing; patent search and

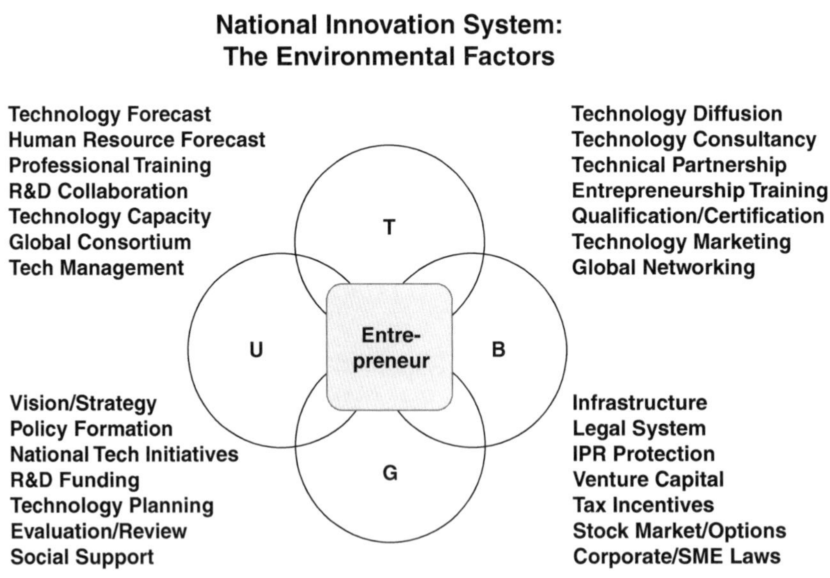

Figure 9. The Innovation Eco-System Factors

analysis; intellectual property rights management; financial planning; laws to facilitate business formation and restructuring; tax incentives and support for industrial R&D; venture capital; and technology clusters. Entrepreneurship is a risk-taking affair. The aim here is to provide support to the entrepreneurs at the firm level in such forms as mentorship, consultation on business models and business plans, guidance in entrepreneurial culture, etc. These elements are shown in Figure 9.

Chapter 7

ROLE OF THE UNIVERSITY

Conventional wisdom tends to credit the university as being the source of innovation. This is a misconception. Innovation can be initiated from any point by anyone in the national innovation system. The university has no monopoly on innovation. Thomas Edison, who held over 1,000 U.S. Patents in various types of electrical devices and applications, pursued his work quite independent of any university. Wilbur and Orville Wright, who pioneered the flying machine and laid the foundation for air transportation and space exploration of modern times, never attended any university. Steve Jobs, arguably the most important innovator in the knowledge economy, never finished his college education.

However, the university is the place where most young adults get educated and scientific research gets started. It is distinguished as a collection of brainpower as represented by the faculty, staff and students. It provides an environment where students and professors meet to be stimulated and challenged, to learn cultural values and befriend each other.

I was amused on reading a conversation between Confucius and his student Tzu-Lu who was a smart young lad with a somewhat flamboyant temperament. He liked to challenge his professor. The exchange went like the following:[67]

> *Tse-Lu said, "There is a certain bamboo in the South Mountain whose sticks stay straight without additional stretching. If used as arrow, it hits at the leather without fail. With a natural gift like this, what good is it to study?"*
>
> *Confucius answered, "Uh! If sharpened to a spearhead and stabilized with a feather, wouldn't it pierce much deeper?"*
>
> *Tse-Lu bowed, "Indeed it would. Your instruction is well taken."*

I can feel the excitement of the exchange between professor and student. Also, from this story, I can tell Confucius had a clear view of science and technology and knew the difference between research and development. He understood well the natural characteristics of the bamboo and that, for practical application as an arrow, it would be much more effective to process it further by sharpening and stabilizing. Turning scientific knowledge to value-in-use, this is what innovation is all about.

Generally speaking, the work of the university is teaching, research and service. A university is judged by the graduates it produces and their contribution and leadership in society. Equally important, the university is judged by the quality of its research and the impact of the knowledge it has created. These are the two sides of the word "service." It should be noted that money or financial assets is not, per se, a part of the university's work or performance. Money or financial assets come into the picture only in their ability to support and sustain the university's functions in teaching and research. This has been recognized as the normal mode of operation of the university for hundreds of years.

In the globalization era, the university should educate its students with world perspectives, awareness of the interdependence of people, organizations and nations everywhere in the world. The students need to cultivate mutual respect for people and the habit of not treating others as they would not want to be treated. This is Christianity, Confucianism and the true teaching of many religions.

World-class

Obviously, most universities have set their goal as becoming a "world-class university." But few have agreed on the standards of what would constitute world-class. In my view, a university aspiring to be world-class should aim at recruiting talents worldwide to teach, attracting the best young adults worldwide to study, and providing an ambience for the faculty, staff and students to interact so that the graduates it produces have the ability to work and excel in any part of the world they choose. I believe this is what a "world-class university" is all about.

While focusing on research and teaching as their primary goals, many universities do want to play a more active role in the research commercialization process or, more broadly, in the knowledge transfer process.

After all, this will make the university more relevant to social progress and may also broaden employment prospects for its students.

Duties of the University

If a university is seriously interested in knowledge transfer, it should state so in its charter or mission statement. And, more importantly, it should establish a review and reward system for the faculty to reflect this goal. Of course, this should be weighted in conjunction with the missions of research and teaching. Universities like MIT, Stanford University and Harvard University in the U.S. and Cambridge University and Oxford University in England, are considered among the top world class universities by any standard. They also have outstanding achievements in knowledge transfer. It would appear that these three roles are compatible and complementary.

At present, in most universities, the substantiation and promotion of faculty members is basically conducted on performance in two spheres: research and teaching. Over the years, universities have established many systems for scoring the number of research publications, the quality and possible impacts of the research, and giving credit to teaching efforts and teaching quality. But few universities have a comparable system for evaluating efforts and accomplishments in knowledge transfer where its scope and impact are more difficult to assess. This is why many younger faculty members naturally who care a great deal about substantiation and promotion will, under normal circumstances, not make dedicated efforts regarding knowledge transfer. If such efforts are not properly rewarded in their career development their potential will not be fully utilized and the contribution of the university to the society will be limited.

Another aspect of research commercialization is the issue of time and commitment. Most faculty members assist their graduate students or research associates in pursuing commercialization of a technology in which he/she has expertise. In many cases, the faculty is the source of the research or technology. Under the present system, faculty may only participate in the start-up in a part-time and advisory role, which is actually inadequate and difficult to enforce. It would be to the university's advantage to design a mechanism for the faculty to take professional leave for knowledge transfer, much like sabbatical leave, but on a different frequency and duration.

In this regard, I would say that any faculty member who is making an effort to commercialize a technology that originates from his/her lab should first find a leader for the commercialization effort. Many scientists are reluctant to admit that someone else may be more suitable to be the champion. Self-esteem is human nature. But one has to realize that people have different skillsets, temperaments and priorities in life and no one is Superman. I have hardly ever seen an individual who is a good scientist, good teacher, good entrepreneur and good business executive, all in one person.

Nobel laureate William B. Shockley may serve as an example.[68] Shockley was an outstanding scientist with Bell Laboratories, and was awarded a Nobel Prize in Physics in 1956 for his contribution in semiconductor junction transistors. In the same year he left Bell Labs to set up Shockley Semiconductor Laboratory for manufacturing semiconductor devices. He was known to have a domineering demeanor and was difficult to work with. Soon, eight of his key researchers left to form another company, Fairchild Semiconductor.

Shockley's entrepreneurial efforts ended quite abruptly. It should be noted that Fairchild Semiconductor was later known as the cradle of microelectronics over the next 20 years. Over 65 new enterprises would be formed by people with employment history at Fairchild. These "Fairchildren" have made great contributions to the region that later became known as Silicon Valley. However, many observers have credited Shockley as the one who brought silicon to Silicon Valley.

After his company closed, William Shockley went into academia and became a professor at Stanford University. Over the years, I have seen many excellent ventures fail because the champion lacked the ability or time to execute the business plan. The culture of entrepreneurship is of fundamental importance and the subject is discussed in Chapter 15 of this book. More details on Silicon Valley are in Chapter 16.

Ranking of Top Universities

In the recent two decades, many private organizations have dedicated themselves to ranking universities worldwide using different measurement standards. By reference to the two most popular organizations, The Times Higher Education World University Rankings and the QS World University Rankings, some general observations of the 2015/2016 results can be made.[69]

- The U.S. ranked first in the number of top universities. It has about half of the top 20 best universities and about one third of the top 100 universities.
- The U.K. ranked second.
- The Asia Pacific region has about the same number of top 100 universities as the EU nations.
- For universities under 50 years old, the Asia Pacific region has the top 6 of the 10 best.
- In metropolitan areas, Hong Kong, London, Boston and Singapore have the highest density of top universities.

The achievements during the last two decades in tertiary education in the Asia Pacific region are remarkable. It is not difficult to link the economic performance and innovation capacity of the nations or regions with the strength of higher education. Therefore, education should remain in the top priority list of any forward-looking government.

There is a love-hate dilemma between universities and university rankings. Except for universities in the top echelon, most others do not want, yet cannot refuse, to participate in the ranking. Once the results are published, universities often voice more displeasure than pleasure with the outcome. This is not only due to the rise and fall of their own standings, but also the relative performance of rival universities, in the eyes of the donors. The ranking organizations themselves are often the source of debates or disputes since the methodologies and standards are often arbitrary and subjective.

For example, the measure of research quality heavily relies on citation indices of publications. This has sparked a numbers game for some authors, and indeed institutions, to manage their number of publications. The pressure to publish has led to cheating or faking of research results and/or authorship, becoming a matter of academic integrity.

Another gray area is the research funding source which can be diverse and complicated and not always comparable in different universities. Yet another glaring example is the assignment of scores based on the number of memberships in the Academy of Science or Engineering in the faculty. In some countries, such membership often reflects more the political or social status of the individual than actual research or teaching excellence. In some places the title represents a license to get government fundings.

It was not uncommon to see world-class journals revoking scientific publications due to the dishonesty of the authors or co-authors, including senior members of institutions or members of academies.

In any case, rankings can be useful for the university as a benchmark of development, but it should be focused more at the academic school/college level and garner more inputs from professional/technical communities. This bottom-up approach can help a university in setting targets toward becoming a "world-class university."

Chapter 8

ROLE OF BUSINESS

The business community, in the broad sense, consists of the manufacturing industry, service industry, wholesale and retail sales, trading and commerce, banking and financing, logistics and transportations, etc. In all these sectors, technology is an important element. It is the mainstay of the economy of a country, a state or any economic entity. The strength of the business sector reflects, to a great extent, the power of any nation. It is little wonder that, after the Industrial Revolution, the gross domestic product (GDP) has been regarded as a primary measure of national strength. The business community, as defined above, contributes the major component of GDP. The GDP per capita, likewise, has been regarded as a measure of the wealth of a country.

On the individual firm level, the rise and fall, loss or profit, of a business is determined by market forces of supply and demand and the skill of its corporate management. Under an open and free market system, a business firm tends to maintain its profitability and competitiveness via innovations: new products, new processes, system integration, continued improvement, and the adoption of proper business models. Innovations help the business run more efficiently and effectively.

In most developed economies the business community is the driver of the national innovation system. As such, it works with the university to broaden the scientific frontier and train students who can readily take up commercial assignments. It advises technology institutes on setting appropriate and timely goals for their projects. And no less importantly, it advises the government to make long-term development goals, set economic and human resources forecasts and build an infrastructure conducive to the growth of business in both hard power and soft power attributes.

A business organization aims to create value for its stakeholders and society and make profit for its investors. Its success is measured by its ability to market products or services, make adequate levels of profit and sustain its existence over time. The type, sophistication and uniqueness of technology is not the measure of success of the business per se, except as a reflection of its marketability, profitability and sustainability.

Because of its direct participation in the market, the business should have a thorough understanding of market needs and make a good appraisal of the relevance of technology to the market. Generally speaking, it should play the role of leader in the national innovation system.

In reality, industrial companies often make mistakes in the commercialization of research results by underestimating the difficulties of promoting new products, especially in an area outside of its core expertise.

The Microwave Oven: Timing, Timing, Timing

A classic example is the microwave oven. The technology was developed by the Raytheon Corporation which was a leader in radar technology during WWII. In 1947, the company patented its high performance magnetron technology aiming for civil applications. The household products market was undoubtedly a growing field but was really foreign to the Raytheon Corporation. Failing to bring any products to the commerical market for a number of years, Raytheon decided to license the technology to Tappan Company of Ohio, an expert in making and selling stoves.

In 1955, Tappan introduced the first wall mounted microwave oven for home use. With a price tag of US$1,295, the equivalent of US$12,000 in 2017, not many units were sold. In 1967, Raytheon acquired Amana Corporation to produce the first countertop microwave oven Radarange with improved technology and brought the cost down to US$500, an equivalent of US$3,600 now. Again, not too many units were sold. Market reluctance and sluggish sales were also caused by consumer worry that radiation might escape to cause hazards. This fear was not alleviated until 1971 when new government product safety guidelines went into effect.

However, the novelty of the microwave alone was not enough to make sales. There was not much demand for high performance ovens. Another 10 years came to pass when more housewives held full or part-time employment. Thus, a need for fast preparation of foods emerged. In parallel to this, a number of food companies started to market frozen dinner packages and appliance companies produced refrigerators with larger frozen storage space. These developments set the stage for microwave oven applications. But the price was still too high.

It was not until the mid-1980s when the less expensive Japanese and Korean products came to the U.S. market that the price of the microwave oven was brought down to a user-friendly level. In the 1990s, the microwave oven became truly a household item.

Therefore, the success of the innovation process was determined not by the technology alone but by surrounding social, political and economic factors: government regulation, employment statistics, lifestyle change, public health, foreign competition, etc. The history and the external factors on the development of the microwave oven was recently highlighted by Susan Strasser.[153]

Therefore, the business community misjudges the market dynamics at its own peril. There are other challenges and risks to the industry and business regarding innovation. The irony is that the success of some technology companies has led to their demise. They tend to focus on maximizing their competitiveness among rivals[70] and neglect to observe how the overall social and technological environment has changed in the knowledge-based economy.

Photographic Films: Eastman Kodak

Take Eastman Kodak Company, for example. For nearly a hundred years, it was the best of all photographic film companies in the world. In fact, Kodak was one of the best recognized brand names of corporate U.S. worldwide. To maintain its lead, it also engaged actively in R&D of new materials and processes and continuously brought to the market newer and better photographic film, instant cameras and related products. It was undoubtedly the leader of innovation in the photographic business. But where is the company now? What has gone wrong for Kodak?

In January 2012, Kodak filed for Chapter 11 bankruptcy protection, announced that it would stop making all digital personal cameras and pocket video cameras, and sold its photographic film business.[71] Later it also sold its patents for over US$500 million to a group of companies including Amazon, Apple, Facebook, Google, HTC, Microsoft and

Samsung. It has restructured into another company owned by the UK-based Kodak Pension Plan. By the start of the 21st century Kodak basically disappeared from the market.

In fact, the photographic film and instant camera markets have nearly vanished altogether. This is because advances in digital microelectronics have made traditional photographic films unnecessary for the everyday consumer. Kodak certainly knew about these technology developments but failed to recognize how rapidly they would become integral parts of personal computers and especially smartphones. Unfortunately, Kodak never suspected smartphones would become its competitor, and indeed, terminator. As new technologies are accepted, old technology becomes obsolete almost immediately. Similar situations have overtaken Xerox copiers, Hewlett-Packard calculators and numerous others.

The management of disruptive innovation is a tricky undertaking and very dangerous if not handled properly. Most big corporations in the U.S. or Europe established a substantial professional workforce to set up their R&D department. In these companies, the R&D efforts are usually financed by existing business units. As many existing units see it, supporting disruptive innovations that the R&D guys are working on is nurturing a force that, if successful, will become a terminator of its own business. What a prospect this will be! How enthusiastic will these business units be in supporting the new R&D program? The answer should be fairly obvious. Therefore, disruptive innovations seldom grow from inside a big industrial corporation. As suggested by Peter Drucker, disruptive innovation is a two-edged sword.[72] It is the prime mover of economic progress, as pointed out by Joseph Schumpeter almost a hundred years ago. But the business community needs to manage it for its benefit and not be imperiled by it.

Strategy of Merger and Acquisition: Cisco Systems

Merger and acquisition is an option used by some companies to expand into new and important fields. Since smaller companies and even start-ups can be more efficient in developing novel technology, they can be acquired or merged and further developed into major and sometimes, disruptive innovations. This, in essence, is to drive cooperation in industrial R&D through market forces.

Cisco Systems is an example.[73] Cisco was founded in December 1984 by the husband-wife team of Leonard Bosack and Sandy Lerner, both computer scientists of Stanford University, and three other co-workers. The product they developed were routers that would support multiple network protocols. In other words, they were able to network various local networks miles apart to communicate and maximize computing power. Think about the spillover impacts on the society. What a splendid idea! Although the formation and the funding of the company was somewhat tortuous, the product was met with immediate success. The sales revenue climbed at almost exponential rates. Very soon, the Bosack couple did not need to cover company expenses with their personal credit cards.

To keep up with market demands and rapidly changing technology, the company quickly transformed itself to become a service provider. Knowing that its technology base was too narrow and the timing too short to develop all the necessary technology internally, it decided to pursue a strategy of merger and acquisition for innovations. As an example, in the one year period of 1995–1996, the company completed 11 acquisitions. Most were small companies with the specific niche technology or products that could be quickly modified and integrated into Cisco's own system of products. This helped build up Cisco's routing and switching capability to service systems of different architecture and intelligence.[73] The company went public six years later, listed in NASDAQ with a market capitalization of US$225 million.

The growth of the internet in the mid to late-1990s quickly changed the IT landscape. Multiple protocol routing was gradually giving way to internet protocol. Pursuing a similar strategy, Cisco was able to stay on top of the internet wave and gained a dominant position as an Internet Service Provider (ISP). At the height of the dot-com bubble of the year 2000, Cisco became the most valuable company in the world with a market capitalization of more than US$500 billion. For the next 15 years it remained one of the world's most valuable companies. The Cisco success story started first with technology push, followed by market pull and was consummated by merger acquisition to sustain its market leadership.

Its acquisition strategy helped Cisco to move forward into the cloud computing age. With the continuous acquisition of new technology companies, Cisco quickly created management functions for system integration and improved efficiency.

In April 2014, Cisco Systems announced that it would establish a US$150 million fund to support early stage start-up firms globally to focus on the Internet of Everything (IoE). The fund was dedicated to IoE accelerators and start-ups. Perhaps this is a new approach to allow more freedom and greater latitude in the future development of innovations.

Building New Bases for the Future

For a business to take on leadership in driving innovation, it has to learn how it can take advantage of new technologies to help its own growth. Indeed, companies such as Intel, Google and Amazon are making major efforts to nurture new dimensions. They have attempted to branch off to new fields or spin off new ventures that might potentially be disruptive to their mainstream product lines. Google has taken up new fields such as self-driving automobiles. It even created its own new mother company, the Alphabet, which can move into rival or competitive areas.[74] Amazon's CEO, Jeff Bezos, has made major investments to acquire the *Washington Post*, and the high-end supermarket chain Whole Foods Market, which can be valuable in the development of future e-commerce.[75] Hewlett-Packard has announced it will split into two companies: one focused on business services and the other on personal computer accessories.[76] These are examples of how technology companies have managed to stay innovative and hopefully achieve long-term sustainability.

Another instrument for the business sector to achieve sustainability is to strive for more substantial partnerships with universities and technology institutes. It is this kind of synergistic interaction that has given vitality to Silicon Valley, making it the most successful eco-system in innovation from the middle of the 1950s to the present. Chapter 16 of this book is dedicated to the growth and legacy of Silicon Valley.

Chapter 9

ROLE OF THE GOVERNMENT I

Role of the Government: The Basics

The government's role is to provide an eco-system or environment conducive to the development of science, technology, innovation and entrepreneurship. The government is the planner, enabler, facilitator and orchestrator. The actions that government will undertake fall into two categories: fundamental and functional.

The fundamental actions include: forging a national vision or national thrust objectives; promulgating strategies and policies to achieve the vision and objectives; setting up the framework of law and taxation; and rallying social support. The functional actions are specific measures taken in certain policy areas including regional planning, physical infrastructure, human resources development, funding support for R&D, protection of small and medium-sized enterprises (SME), protection of intellectual property rights (IPR), developing venture capital funds and a stock exchange for technology start-ups, and enhancing capability in project management.

Fundamental Roles of the Government

A national vision is fundamental because it provides cohesion for all the parties and a focus for their efforts. A good example is U.S. President John Kennedy's 1960 decision to launch the Apollo Program.[77] He said, "In ten years, we will put a man on the moon and bring him back safely." This was a heroic, romantic and inspirational goal, and the nation rallied to it.

To fulfill a national vision, proper strategies and policies need to be developed. For the Apollo Program, the U.S. Congress appropriated an immense budget for its implementation. The Administration rolled out programs spreading over different government agencies in commerce, energy, education, defense and institutions such as the National Science Foundation (NSF), major universities and technology corporations. It mandated the National Aeronautics and Space Administration (NASA) to coordinate and orchestrate various projects and take charge of mission control.

The vision was fulfilled in less than 10 years. It did not create any quantifiable financial benefits with regard to the real estate on the moon. But it had bolstered the development of computer science, space science and engineering, microelectronics, telecommunication, advanced materials, advanced combustion, sensor technology, precision machinery, etc. The research results have gradually undergone technology diffusion in the past 50 years. Many innovations and their derivatives have flourished in the market economy. The accomplishments in these fields have placed the U.S. solidly in the lead in the world development of science, technology, innovation and economy in the 20th century. It also prepared the U.S. to leap quickly into the knowledge economy for the 21st century.

This was the largest scale technology effort of the U.S. in the post-WWII period. Participants in this program often recalled the team effort, entrepreneurial spirit, and strong feelings of dedication and honor.

A similar example, less publicized and in fact more secretive, was the decision of Chairman Mao Zedong in late 1957 to launch a satellite capable of orbiting the earth. A great deal of technology development and preparation started under very difficult resource constraints. The program was accomplished in 1970 when China was able to successfully launch its low orbit satellite East is Red (東方紅). This enabled China to later establish its nuclear defense capability. And, because of its proclaimed national status, albeit only known at highest government level, it was allowed to remain relatively unscathed by the 1965 political rampage of the Cultural Revolution which set the country back economically for at least 10 years. The program, referred to as Two Bombs-One Star (兩彈一星), laid the foundation for many innovations in later decades.[78] Unknown at the time, due to the overall political structure and the restrictive national innovation system, the program's spillover benefits to China's economy was delayed until nearly 30 years later.

Back in the U.S., decisions of Presidents Gerald Ford and Jimmy Carter in the 1970–1980s to fight cancer and President Ronald Reagan in the 1980s to control AIDS can be viewed within the framework of setting a national vision. Substantial funding support was appropriated for R&D programs in related areas. Viewing it from this perspective, while these two diseases are still prevalent, definite progress has been made in their understanding and treatment. These have also given the U.S. a leading edge in biochemistry, biophysics, genetic engineering research, and related innovation and entrepreneurship efforts.

Many efforts can be analyzed from this perspective. During the 1980s, leading East Asian economies, Singapore, South Korea and Taiwan, pursued economic development with concerted national efforts.[79–81a] Since the available resources were limited and competitive pressure high, they each had to select certain subjects as thrust objectives for development.

National Thrust

Generally speaking, the selection of national thrusts can be based on the following four categories:

A. To achieve a national vision or certain high level objectives.
B. To find solutions to specific economic or social problems.
C. To enhance productivity or competitiveness.
D. To broaden frontiers of knowledge or create new opportunities.

The Moon Landing Program and Space Satellite Program described above belong to Category A. The objectives were specific and the resources needed huge and prolonged.

Programs towards control of cancer and AIDS belong to Category B. Other examples in this category may be programs to control epidemics, solve environmental pollution in the air, water and on land, reduce carbon dioxide emissions or prevent natural and industrial hazards.

Category C covers a wide variety of programs in many branches of science, engineering and technology. It includes: automation, advanced chemicals, advanced materials and processing, robotics and precision machinery, micro-electronics and nano-electronics, digital manufacturing process, sensors and sensor technology, telecommunications, internet technology, energy conversion and energy storage, power electronics, genetic engineering, quality engineering and management, etc. These are areas important to national productivity and competitiveness and were taken up as objectives of development in many countries. However, since they are highly competitive, any further development should be linked closely to existing industry and industrial needs. Thus, a careful Strength Weakness Opportunity Threat (SWOT) analysis is recommended.

Category D also covers a wide variety including: cloud computing, big data and big data applications, bio-informatics, advanced energy systems, global satellite navigation network, personal medicine, artificial intelligence, 3D printing, virtual reality, etc. Also, selection of these objectives should be guided by detailed analysis and technology and human resource forecasts.

For many developing economies, governments were eager to take initiatives to generate such analyses. With the participation of the institutional players from the innovation system and, occasionally, external professional consultation, the government would establish certain objectives for national development. It would follow up with R&D funding support, tax incentives, and legal reform to promote technology businesses when the project had been developed to reach the commercialization stage. This general approach was effective in Singapore, South Korea and Taiwan over a 30-year period.

However, some economists have faulted this approach as equivalent to picking horses randomly in a horse race and thus contradictory to market economy principles. My view is that the market economy is important for innovations to flourish and create social benefits. But it does not mean that the government should sit on its hands and do nothing. Each government has an obligation to find the best possible approach to facilitate solutions for its country especially when competition is fierce and resources limited.

But visions or thrust objectives would just remain dreams if no implementation strategy or policy measures are provided to make them happen. Formulation of strategy and policy requires not only the consideration of theoretical context but also the reality under specific time and space. There are no set rules on how these work except that they should be formulated with full participation and thorough discussion on the part of the institutional players, U, T and B under the leadership of the government G.

The CEO of a major city once asked me how many new jobs the government's proposed science park would create by the following year. He was astonished to hear that it would depend on what efforts the government made to empower technology and business development in the park.

And in any case it would be better to think about it in a time frame of five to ten years rather than one or two.

Setting the rules, making them transparent, and applying them fairly to all players are of fundamental importance in any game. In a civil and democratic society, this is based on the rule of law. The law will cover business endeavors and management functions, company establishment and closure, business operation and transaction, taxation and distribution of profits, etc., and will apply equally to everyone.

In the past, when I visited a city or a country, I usually inquired about the procedure and length of time it took to open and close a company. I was often taken aback when government leaders replied that the rules were complex but they would be flexible and give each case special consideration. I invariably advised them that hardworking entrepreneurs would appreciate consistency and transparency of government codes and equal accessibility much more than random flexibility or personal favor.

The Functional Roles of the Government

What should governments do to facilitate the growth of the knowledge economy? Figure 10 below illustrates the functions that government can and should take.

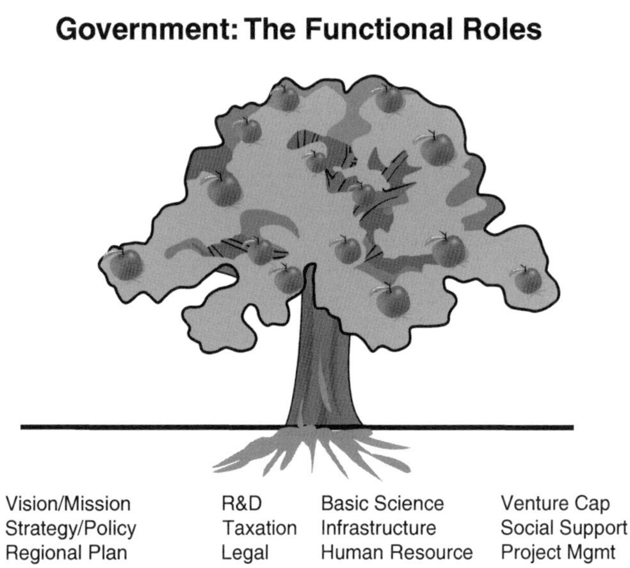

Figure 10. The Roles of the Government

These functions are depicted as the roots of the knowledge economy. Except for building physical infrastructure and promoting basic science, all other roots are related to the political and social fabric of a country. They are soft power related attributes, namely: national vision and mission, national strategy and implementation plan, regional planning, R&D, simple and fair taxation scheme, just and transparent legal system, human resources, venture capitals, and, capability of project management. These are now well known and subjects of much study. I will not make further comments in this book except on cases where my personal observations may be of specific interest. It should be emphasized, however, that to perform its role, political stability of the government is of fundamental importance.

Chapter 10

ROLE OF THE GOVERNMENT II

Role of the Government: Too Slow or Too Fast

The consequences of government inaction, or inadequate action, are serious. The nation will sooner or later be left behind in the era of competition. It is not only governments in the third world that have this problem; developed economies can also run into this situation because of complacency. The challenge encountered by Eastman Kodak can also occur in government.

Hong Kong: Past Glory

Take Hong Kong as an example. In the last century, Hong Kong enjoyed a unique geographical advantage as the gateway between China and the rest of the world. By offering efficient financial services and a stable and lawful environment for trade and commerce, it gained substantial prosperity and world competitiveness.

After 1997 the environment underwent rapid changes. Many cities in China, especially Beijing, Shanghai, Guangzhou and Shenzhen, have worked extremely diligently in developing technology to bolster economic strength. They also built direct connections and relationships with the world. Hong Kong's traditional advantage as the entrepot has rapidly eroded. In addition, the Pearl River Delta (PRD), which previously relied on Hong Kong as the source of capital and technology for its manufacturing business, has nurtured remarkable technical and business capacity. The four Chinese cities above have attained double digit GDP growth over the past 20 years. They all have become metropolis with worldwide connections in tourism, travel, trade and industrial supply network.

However, Hong Kong's leadership continued to indulge in the "Asia's World City" or "Super Connector" fantasy. The *laissez faire* mindset remained prevalent not unlike colonial times. It was passive even to private initiatives in promoting substantive collaboration between Hong Kong and the PRD.

A case in point is the effort of HKUST when it spearheaded a program in 1999 to build the Nansha IT Park near a historical site midpoint between Hong Kong and Guangzhou. The park was intended to be a base of education, knowledge transfer and technology entrepreneurship linking Hong Kong and PRD. It will enable the results of scientific research from Hong Kong to be applied in the PRD with the participation of local talents and resources. The program has the full financial backing of the Fok Ying-tong Foundation, policy supports of Guangzhou municipal government and the State Ministry of Science and Technology. But HKSAR government has been skeptical and aloft. Today, Nansha has become a highlight of the 13th Five Development Plan of Guangdong. The harbor capacity for cargo container shipping of Guangzhou-Nansha has become the sixth largest in China and is expected to overtake Hong Kong as the fifth largest in a year or two.

Hong Kong's economy has been lacklustre and fell much behind the neighbouring Guangzhou and Shenzhen. In 1988, Hong Kong accounted for 18% of China's GDP. In 2015, it dwindled to about 2%.[156] The relative inaction of the Hong Kong government left it marginalized. Recall the proverb: "Like sailing in upstream water, to stand still is to be left behind."

The ineptitude of the government in the last 15 years has heightened social discontent and sharpened the discord among the youth and the underprivileged. The 2014 Occupy Central Movement, though misguided and unlawful, that lasted for 79 days was fundamentally due to distrust and frustration of the post-1980s and post-1990s generation.[168]

Crisis of Vision: Invest for Future

Hong Kong is now in a soul-searching mode, trying to formulate a vision and policy for the future. Actually the answer is quite obvious.

Back in 1997, a MIT report "Made by Hong Kong" had outlined the challenges and opportunities for Hong Kong. It called for the development of technology and innovation.[155] For a long time, the academia circle has argued

Table 1. R&D and Educational Investments of Hong Kong in 2013

Hong Kong's Investments on R&D and Education as % GDP in Year 2013			
A. Similar GDP (in Billion US$)			
		R&D	Edu
— Denmark	330.82	2.98	7.88
— Singapore	295.74	2.04	2.96
— Israel	291.57	3.93	7.95
— Hong Kong	273.66	0.73	3.76
— Finland	256.84	3.55	6.35
B. Similar GDP/Capita (in Thousand US$)			
— U.K.	39,351	1.73	6.05
— Japan	38,521	3.35	3.54
— Hong Kong	38,071	0.73	3.76
— Israel	35,845	3.93	7.95
— Italy	34,703	1.27	4.16

Source: IMD World Competitiveness Yearbook in 2014

that the government should focus on strengthening human resources and technology capability to broaden its economic base and competitiveness.[82a,82b,82c] One clear action was to increase investment in total R&D and build an environment conducive to innovation and entrepreneurship.

The inadequate total investment in research and innovation in Hong Kong is unmistakable. Shown in Table 1 is a comparison of investments on R&D and on education of several countries comparable in economic size (GDP) or in wealth (GDP per capita).

It is abundantly clear that while Hong Kong's investment on education has been at comparable levels with these countries, Hong Kong's investment in total R&D was far off the mark. In fact, Hong Kong's investment in R&D was like a third world country's. It should be noted that all the data were taken from the IMD World Competitiveness Yearbook which has set clear definition and consistency in the selection of data.

It appeared incredible that while all developed economies have dedicated a significant amount of resources in R&D to build up its technology and innovation capacity, the Hong Kong government did not view investment in R&D, an investment for the future as a priority.

It was especially puzzling since Hong Kong was not in any budget pressure. For the past 12 years, Hong Kong has experienced an annual budget surplus. It now has cumulated a HKD 908.3 billion cash reserve.[157] But the government has resolved to do the minimal in nurturing technology and innovation. The efforts to encourage industrial R&D, promote industry-university collaboration, and support technology entrepreneurship were simply dismal.

As a reference, for the year 2013, total R&D expenditure in percentage GDP of Shenzhen, Taiwan, Singapore and Hong Kong are 4.03, 2.99, 2.00 and 0.73, respectively (see also Figure 18 in Chapter 18).

It is little wonder that in 1988, Shenzhen strived on fishing and agriculture but today, 50% of its GDP is from technology. Of course, during this period, Hong Kong government has boasted to be the super connector all the same.

Building on Strength: Higher Education

A recent study on the future development of Hong Kong and the PRD has been jointly undertaken by the Chinese Academy of Engineering (CAE) and the Hong Kong Academy of Engineering Sciences (HKAES). It pointed out that while the city was gradually losing its traditional advantage, it had accumulated impressive strength in its institutions of higher learning.[163]

After 1997, a wave of talented professionals of Chinese origin who had previously worked in America and Europe were attracted to the new prospects offered by Hong Kong's closer ties with their homeland. They gathered around the universities doing what they do best: research and teaching. Five universities in Hong Kong have gained world reputations in the last decade. This was demonstrated by the top world university rankings as shown in Figure 11.

The CAE-HKAES Joint Study thus suggested that the future Chief Executive should budget an investment of 3% of GDP for R&D over a five year period. This would take full advantage of the high quality professionals in the city and its experienced business community. Apart from supporting university research in basic science and engineering, government should assert its leadership in strengthening innovation capability, promoting collaboration between the university and industry and re-doubling its

Globally Recognized Universities

(Source: CAE-HKAES Joint Study Report, April 2017)

Rankings	**HK**	**London**	**Boston**
City having the greatest no. of world's Top 100 Universities by QS	#1 (4 out of 8)	#1 (4)	#2 (3)

Rankings	HKUST	HKU	CUHK	CityU	HKPolyU
World's Top 150 Universities in QS World University Rankings 2016	#36	#27	#44	#55	#111
Global Employability University Ranking 2016 (reported by Times Higher Education)	#13 (#1 in China)	#77	#85	#138	–
World's Top 65 Universities in Engineering & Technology • QS World University Ranking 2015	#14	#26	#41	#60	#44
• Times Higher Education Rankings 2016	#19	#33	#74	#58	#74
World's Top 50 Universities in Computer Science & Information Systems • QS World University Rankings 2016	#14 (#1 in China, #2 in Asia)	#19	#22	#49	#49

Figure 11. World Ranking of Universities in Hong Kong

Note: HKUST — Hong Kong University of Science and Technology
HKU — University of Hong Kong
CUHK — Chinese University of Hong Kong
CityU — City University of Hong Kong
HKPolyU — The Hong Kong Polytechnic University

support for entrepreneurship. Additionally, the study suggested, the new Chief Executive of Hong Kong SAR should establish a high level advisory group for science and innovation. This group would provide guidance for the Chief Executive to rise above the bureaucratic cobwebs and forge a vision for the future of Hong Kong.

China — Phenomenal Growth

It should be noted, however, overzealous action of government can also unknowingly create problems for the future. Many governments are apt to take expedient actions for quick results rather than gradual, systematic nurturing of the roots of the economy tree (Figure 10). Noting the success of Singapore, South Korea and Taiwan in economic transformation during the last quarter of the 20th century, many cities in China jumped on the bandwagon of manufacturing as the vehicle of economic growth.

Without establishing a strategic planning process to forge a vision and regional plan, many cities quickly set up their own policy measures to attract foreign investors, including those from Hong Kong and Taiwan. They laid out attractive incentives to bring in capital and technology. The major tool of local governments was, quite naturally, the land.

In China, usage of the land is under central planning and control of the State. By leveraging the prerogative of granting "the right to use" as a tool, lucrative incentives were given to investors to lease the land for industrial uses. Local governments were also generous with seed investments to build up the physical infrastructure. The land was generally flattened and provided with electricity, gas, water, telecommunications and transportation to expedite the operations of factories and trade facilities. This was generally referred to as Five Flows-One Flat (五通一平). Thus the economic development has enabled many local governments, through their financial instruments, to own a share, and sometimes the lion's share, of the technology businesses under their jurisdiction.

Knowing that they need professional knowledge in developing the technology business, some local governments also set up technology institutes on their own or in partnership with external academic institutions. These actions sent welcome signals to national academies, research centers, institutes and universities. The technology institutions, with liberal interpretations of their own mission, objectives and capability, were eager to leverage various government efforts for their own growth.

This was how the government quickly became a participant in commercial development. Likewise, university or technological institutes all have an interest in the commercial marketplace and want to quickly create business revenue as an organizational objective. This was one of the reasons that China, a leading socialist country under the pretext of a planned economy, became highly flexible and capitalistic in the market economy in a period of 30 years.

Additionally, in the process of "privatizing the state-owned enterprises (SOEs)," especially those with monopoly rights on minerals, utilities and natural resources, the governments tended to release a part of the ownership to private investors for the cash it needed. In the process they allowed themselves, as the major shareholders, to hold on to management. Thus, while the SOEs became privatized, the government still technically owned the businesses. Better yet, the government could appoint the chairman or

CEO of the enterprise and have its hands firmly on the management. Thus it has enjoyed the best of both worlds.

Over a 30-year period, the Chinese economy grew over 10% annually. In 2014, the GDP of China overtook that of Great Britain, Germany and Japan to be the world's number two. It has become the world's largest manufacturing base. It is indeed expected to catch up with the U.S. for the number one position in another 30 years, or less. China's rapid rise to economic power in the past 30 years can be said to be a miracle in history.

All Eyes Are On

The 21st century is the era of globalization. China's economic strength has opened its way to the world arena. In the past two centuries China was basically neglected, sidelined and marginalized in world affairs. Few countries would listen to China's view or address the issues it raised. With its new-found strength and assertive attitude, China has been playing an increasingly prominent role in the world community. This has alarmed many nations including the U.S., which traditionally enjoyed an unquestioned leadership position in the Asia Pacific region.

In addition to its strength in the manufacturing industry, China has scored important successes in several large-scale national technology programs. The programs were initiated and managed by the government and supported with national resources. This has included space exploration and high-speed rail.

China's position in outer space is arguably second only to the U.S. It was noted that on April 2, 2017, China's cargo spaceship successfully docked with the Tiangong 2 space station. It was reported that the spaceship carried six tonnes of supplies and two tonnes of fuel. It will enable China to establish a permanent presence in the Earth's orbit.[159] This is a first in the world.

In high-speed rail transportation (>300 km/hour speed), China is second to none. Today, China has nearly 24,000 kilometers of high-speed rail that carries several hundreds of millions of passengers annually. In the 10-day Spring Holiday (Chinese New Year according to the agriculture calendar) period, over a hundred million passengers rely on the rail system for travel as public highways are choked with smoking automobiles. China is also helping to build high-speed rails in 10 countries including Turkey,

Thailand, Russia, Indonesia, Hungary, Laos, Bangladesh, and possibly, Malaysia, India and the U.S.

The progress of China in high-speed rail (HSR) was one of the most remarkable developments in world technology and innovation. In 1993, rail train service in China averaged under 50 km/hr. Recognizing that this was highly undesirable for national development, the 9th National Five Year Plan in 1996 undertook a "Speed Up" campaign. Both the speed and the system were to be upgraded. Technical assistance of foreign technology providers such as Bombardier, Siemens, Alstom, Kawasaki Heavy Industries and others were procured in parallel to an indigenous national program of improvement. By 2008, China had gained independence on all the major elements of HSR technology which included: design, infrastructure, train, track, bogie, signaling, power supply, building construction, system control, passenger management, etc.[63,160]

This extraordinary development of innovation was led by the China National Railway Corporation (CNRC) which orchestrated various players of the innovation system including: the corporate engineering services, the Chinese Academy of Railway Sciences and its many divisions, Southwest Jiao-tong University, Beijing Jiao-tong University, and other research centers and laboratories. Since 2009, The Hong Kong Polytechnic University also joined the effort by making contribution through its state-of-the-art health and safety monitoring technology (HSMT) in fiber optics, structural safety sensing and smart rail operation (program leaders: H. Y. Tam, Yi-qing Ni, S. L. Ho and K. K. Lee) in conjunction with China Southwest Jiao-tong University Railway Development Corporation (CSRD) led by Pang Xiang Wang. Several components of the HSMT were in place in the operation of the Beijing-Shanghai HSR Line and elsewhere. A review of the remarkable cross-border collaboration was the subject of several workshops co-chaired by Alex P. K. Wai (HKPolyU) and Guo-tang Zhao (CNRC).[191] I believe the multi-faceted impacts of HSR in the economic growth and environmental sustainability of the 21st century society will extend beyond the original scope of J. A. Schumpeter.

Perils of Overheating

There are wide-ranging internal problems behind the economic progress. Without adequate strategic regional planning, there was little distinction between the provinces. They all delved into the manufacturing of electricals, electronics, chemicals and consumer products, competing for the same export markets. The differences between provinces are more in size than in kind. Competitive advantage came from low-cost labor, cheap energy and depletion of natural resources. This was not conducive to the development of a well-functioning innovation system.[83, 132]

The rapid increase in the number of automobiles has cluttered city streets and choked the environment. Pollution of the air, water and land, and industrial hazards and product safety issues were large-scale and widespread.

A special trend after 2008 was the central government's focus on developing green energy. In response to this, many provinces have devoted heavy investment to the development of electric cars. Lucrative subsidies have been given to projects to manufacture electric cars, without adequate consideration of the availability of technology and demand of the market. There are repetitions of projects, leading to overcapacity, oversupply and industrial inefficiency.[84]

The rapid increase in manufacturing jobs has attracted a large part of the rural population to migrate to the cities as labor. The basic infrastructure of the city, generally outdated and slow to upgrade, is fragile and stretched to its limits. The newcomers, like new immigrants, find difficulty in integration within the new environment, which has gradually turned unfriendly, if not hostile. In the meantime, housing demands have promoted the springing up of sub-standard dwellings on the one hand and overpriced luxurious ones on the other. This is further fueled by overzealous investors. Overall, social injustice has become a growing area of concern.

Thus the phenomenal riches accumulated in China have short-circuited healthy growth of social institutions and the national innovation system. In 30 years the innovation system became skewed and distorted. Government's influence has overwhelmed all three other institutional players, which became confused regarding their identity, objectives and positioning.

Also, as a consequence of the length of time taken to build up the legal system and judicial culture, corruption became a serious problem in many governmental agencies and institutions. The technological achievements of most universities and institutes remain mediocre. Most state-related businesses have encountered problems of competitiveness and sustainability caused by distorted organizational and structural issues. This is detrimental to the healthy development and sustainability of the society and the economy.

Mid-course Adjustment

These problems surfaced in the last decade. After taking office in 2013, President Xi Jinping launched a major national program to weed out corruption in the provinces, SOEs, ministries and the military, all the way to the top leadership of the politburo. Xi has reiterated the importance of holding on to ethical values and the moral high ground.[85] He vowed to affirm the constitutional rights of individual citizens and private enterprises and establish the rule of law as a basis for promoting innovation and entrepreneurship.

Globally, the government has embarked on the "Belt and Road Initiatives" (BARI) program, intended to help build infrastructure and promote trade and exchange between China and the world.[86] It is intended to create partnership with over 65 countries in Asia, Africa and Europe, and improve the livelihoods of people along the way. This ambitious initiative is in line with the traditional Chinese value of "world harmony." And, in certain aspects, is to avert a direct confrontation with the U.S. which has taken the position that South East Pacific is within its core interest. To help finance the building of these infrastructure, China has led the establishment of the Asian Infrastructure Investment Bank (AIIB) with 52 member states. The AIIB has been open for business since January 2016.[161] The BARI programs have been gradually set into motion. The success of this ambitious program, quite unprecedented in recent history, will depend on the ability to build trust and consensus among the national partners involved. It will require efficient and effective governments, China and its many partners, to play a leadership role along the way. To gain the support of the partnering countries, China will need to make substantial progress in developing its own knowledge economy and green energy.[132]

Overall, the economic performance of China in the last 30 years has been a phenomenal success. It has lifted more than half a billion people out of poverty, an outstanding accomplishment in human history. However, the economy was developed on a course laden with short cuts and expediency. It would require the establishment of a strong national innovation system for its sustainability.

Not the least in sustainability is the establishment and practice of law. Pundits have observed that over the past 70 years since Mao Zedong came to power, China has moved from a 30-year period of "Rule of Politics" under Mao, to a 30-year period of "Rule of Economics" under Deng Xiao-ping, and now into a new period of "Rule of Law" under Xi Jinping.[164] This should be a welcome change in the history of China.

Rise of New Entrepreneurship

It should be noted that a new wave of entrepreneurship run essentially by individuals in the private sector has emerged in the last decade. These entrepreneurs have recognized the opportunities created by internet technology and the globalization era. They have been pursuing the market with special niches of internet innovation coupled with novel business models. While familiar with the domestic culture and environment, they are able to drive their business by taking advantage of standard government policy measures that strongly favor new technology businesses. They also became pioneers in their fields in the domestic market and quickly moved to pursue similar markets in other regions and the world.

These companies include Alibaba, Baidu, Huawei, Tencent and others. While maintaining contacts with local universities or institutes, they strived to lead the development of new technology business through their own efforts. Their successes have signified the coming of a new generation of Chinese innovation and entrepreneurship. Alibaba and Tencent will be further discussed in Chapters 25 and 26 of this book.

Two additional factors that are incidental but gratuitously favorable to the development of technology entrepreneurs should be noted.

a. The vacuum created by rapidly moving technology and the sluggish bureaucracy is working to the favor of entrepreneurs. This has given start-ups the freedom to move according to market rules with relatively little government interference. Luckily, these movements did not leave the government in any negative position since tax revenue and the tax base have been continually growing.
b. The demographic changes in the society has created eager acceptance of new technology products and services. The domestic market has been greatly enhanced by the rising economic strength of the post-1980s and post-1990s generations which have the will and the resources to spend. They not only drive the market internally but also provide momentum to the market globally.

In summary, the government's role in the national innovation system is in setting up national vision and objectives, formulation of strategy and policy measures, and building up infrastructure, with both hard and soft attributes. The government is responsible for establishing the overall environment to stimulate and empower all other players including individual entrepreneurs, to function to their fullest potential. It should not over-extend its functional ability nor underestimate its legal responsibility. It should refrain from participating in the innovation process in a way that competes with or displaces other players. In economic and social development, the government should not sit still but it should not overreach either.

New Developments

There are new recent developments in both China and Hong Kong that should be highlighted.

- **Hong Kong**

On October 12, 2017, the new Chief Executive of HKSAR, Carrie Y.N. Lam-Cheng, said in her first policy address that the government has determined to boost the development of emerging industries in addition to the traditional industries, and considers that both innovation and technology (I&T) and the creative industries have a competitive edge and much potential. She added, "They will not only bolster economic growth but also create quality employment opportunities for the young people."

In order to catch up in the I&T race, Lam-Cheng stated a goal to double the gross domestic expenditure on R&D, as percentage of the GDP, from the current 0.73% to 1.5% within her five year tenure. She will also press ahead I&T development at full steam with other related programs.[180] I think the decision to double R&D expenditure to 1.5% in five years is admirable but too conservative, a far cry from the "full steam" as stated.

It is of interest to note that Singapore was the first international city that Lam-Cheng visited in two months after taking office. One distinguishing feature that should be learned from Singapore is its strenuous effort in building innovation capacity to broaden its economic base. Singapore's gross R&D expenditure as a percentage of GDP in 1990 was

0.85%. In 2004, it jumped to 2.24% and stayed at this level all through this day (as described in Figure 18, Chapter 18 of this book). The efforts of the neighboring Shenzhen and Guangzhou were even more aggressive.

A 2017/2018 global survey on the competitiveness of world cities has just been released. The United Nations, in conjunction with the Chinese Academy of Social Sciences, reported that Shenzhen rose rapidly to 6th while Hong Kong slipped to 12th place.[189] In the same survey, Shanghai, Guangzhou and Beijing took 14th, 15th and 20th place respectively, and were not far behind. This survey considers the overall competitiveness of cities including population density, housing cost, business environment, technology index, university index, income per capita, environment quality, living quality, etc. This is another alarming signal for Hong Kong regarding its competitive status among its peers.

The Chief Executive's determination on boosting the investment of R&D is, nevertheless, an encouraging sign that the new HKSAR government has now recognized the urgent need in diversifying the economy and giving hope to the young generation. This may well be the new beginning for Hong Kong's strive for long-term sustainability. How much Lam-Cheng can accomplish on pressing I&T, and entrepreneurship, will be dependent upon her wisdom and action in lifting the administration from bureaucracy and outdated mindsets.

- **China**

During the week of October 19–26, the world witnessed the opening and closing of the 19th National Congress of the Chinese Communist Party (CCP). President Xi Jinping has reaffirmed many policy measures put forth in the past five years. These include: keeping up with the "irreversible" momentum of the anti-corruption campaign, maintaining the rule of law, protecting public interest, protecting national sovereignty, nurturing friendship and harmony around the world, upholding the "one country–two system" concept for Hong Kong and Macau, opposing resolutely any independence movement of Taiwan, and, above all, striving to make China a nation of leading global influence by 2050. He outlined his vision to make national rejuvenation of China as the focus of the "China Dream" which will be a common bond for all the people.[181]

It was apparent that Xi has received overwhelming support from a wide spectrum of the party membership and the nation. His accomplishments in the anti-corruption efforts and in reviving genuine Chinese culture and values were widely commended. Xi was unanimously re-elected as Party General Secretary of CCP and the President of China. I view his continued leadership with high hope that an era of the "Rule of Law" may be prevalent for the future China which will see the upholding of its soft power as the nation grows in hard power. This will be the essence of the rejuvenation of China. But he should not underestimate the challenges ahead from "Me First" foreign powers, hardcore self-interest groups and a growing young and restless constituent in the society.

Chapter 11

ROLE OF THE TECHNOLOGY INSTITUTE

Secret Ingredient of National Capacity

Technology institutes are an integral part of the technological infrastructure of any country. They exist in different forms in different countries. Generally speaking, they represent a body of knowledge for providing advice and guidance to the government on technological issues relevant to national needs.

Some institutes have physical infrastructure and are real entities employing scientists and engineers to carry out missions. Some exist basically as virtual institutions: they are honorary institutions and rely on the constituent members and their affiliated organizations for actual technical expertise. While some institutes are part of the government, most are non-government organizations (NGOs) that may have varying degrees of government affiliation.

The institutes may not be as well known to the public as universities or colleges because their work is technically focused and they generally interact only with limited sectors of society. This is especially so in developed countries where the universities and industries are strong and have prominent roles that are evident to the general public.

The role of the institute is, more often than not, misunderstood. Countries have established institutes for various reasons and expect them to function differently. Overall, the institute should recognize its unique position and objective in the national system and not become competition to the university or business.

In the knowledge economy, most institutes are concerned with linking scientific research and industrial technology and bringing about value for the public good. They serve as the bridge, connector, catalyst and facilitator for the institutional players. They should especially be a supporter, steward and mentor to individual entrepreneurs.

Most developed countries have a defined portion of its technical capability residing in the technology institute, some with substantial resources and potential. Even in the U.S., whose universities and technology companies are proven leaders in the world of innovation, technology institutes are playing important roles in the national innovation system.

United States

The U.S. National Academy of Science (NAS), National Academy of Engineering (NAE) and National Academy of Medicine (NAM) are institutes of high reputation. They are non-governmental but chartered by the U.S. Congress to function for the public good. They do not operate any research laboratories but jointly administer the National Research Council (NRC). The NRC coordinates policy formation to encourage the use of scientific research in the development of industries and promotion of national security and welfare.[87] It conducts studies through scientific meetings and workshops. The expertise of the Academies lies in their memberships, backed by their universities or institutions. The members are nominated by the membership of each Academy. The NAS, founded in 1916, is the oldest of the three.

After WWII, the U.S. government reorganized most of its defense-related technical centers to National Laboratories (NL) dedicated to specific technology fields such as new materials, green energy, energy conservation and nuclear power safety. The NLs receive funding from relevant federal agencies but are structurally outside of the federal government. The government contracts outside organizations for their management. For example, Argonne NL was managed by the University of Chicago, Lawrence Livermore NL and Lawrence Berkeley NL by UC-Berkeley, Oak Ridge NL and Brookhaven NL by the Battelle Laboratory-Columbus. These NLs each employ over 2,000 scientists and engineers and are major technical institutes in their respective fields. They all have the objective of promoting commercialization of research results.[88]

The National Bureau of Standards in Gaithersburg, Maryland was reorganized to become the National Institute of Science and Technology (NIST) in the mid-1980s. As such, NIST runs many research laboratories in advanced technology fields. It also has a mandate to provide technical assistance to SMEs and entrepreneurs. NIST is organizationally under the Department of Commerce. In addition to its research staff, it also receives visiting scientists from industries and foreign countries.

NIST and the Department of Energy supported National Laboratories, are important players in the U.S. national innovation system. Other U.S. departments also operate technology institutes to assist in achieving their specific missions.

For example, the Department of Health and Human Services operates the National Institute of Health (NIH). NIH has established large research facilities and employed many scientists in Bethesda, Maryland. It conducts internal scientific programs as well as supports other universities in cooperative research. NIH has become the focus of the bio-medical and health science eco-system in the U.S. The Center of Disease Control is another major institute and is instrumental in controlling epidemics and other diseases. The Federal Aviation Agency has contracted the MITRE Corporation as its resource center to develop advanced aviation system technology for improving air traffic control and aviation safety.

On the private sector side there are many technology institutes dedicated to various causes. Many institutes are established as corporations, others as not-for-profit organizations. The Battelle Laboratories headquartered at Columbus, Ohio, for example, is a major institution that operates research laboratories in many technology fields and economics. It also functions as manager of many NLs under contract with the U.S. government.

Interestingly enough, the Bell Laboratories (Bell Labs), established as a part of the American Telephone and Telegraph Company (AT&T) in 1925, was a premiere research lab with worldwide reputation. It had been known to conduct pioneering scientific research in radio astronomy, solid state physics, the transistor, the laser, semiconductors, information theory, operation systems, programming languages and others. In the 20th century, eight Nobel Prizes and two Turing Awards were awarded to scientists for research completed at Bell Labs. The research results of Bell Labs have contributed greatly to the growth of information technology industries benefiting the

U.S. society and economy. Its parent organization, in the course of many business phases (AT&T, Western Electric, Lucent, Alcatel-Lucent), on the other hand, has been unable to capitalize fully on its results to achieve greater commercial success in its areas of interest. In early 2016, Bell Labs was acquired by Nokia and renamed Nokia Bell Labs, much to the chagrin of the U.S. scientific community.

Another institution, the Xerox Palo Alto Research Center (PARC), established by Xerox in 1970, has done excellent research in VLSI, computer systems, laser printing, Ethernet, graphical user interface and computer software technology. Some were instrumental to the development of the modern personal computer. Its contribution to the development of both Apple Computer and Microsoft was well known. But the Xerox Corporation has been unable to turn its results, except with respect to laser printing, into major business innovations for its own benefit. Since 2002, PARC was reorganized to be an independent company conducting advanced technology development in conjunction with many corporate clients and partners.

These examples are indications of the complexity and difficulty of converting good scientific research into good commercialization. This is a subject of much concern even in a technologically advanced country like the U.S. It was no wonder that U.S. President Barack Obama, who took office amid the worst economic downturn of the century, has felt the need for establishing an innovation policy to help reinvigorate the national economy. Promoting the interaction of the university, institute and business sectors was the key policy objective.

Europe and Asia

In the European countries, national or royal academies of sciences and engineering are part of the formal establishment. Most are honorary in nature and exist as a forum of scientific and technological discussion and exchanges. For example, The Royal Swedish Academy of Engineering Sciences (IVA), established in 1919, was the first of its kind in the world and was widely recognized for its advisory role in the economic and social development of Sweden.[89] It has helped a relatively small country with an adverse natural environment to occupy a prominent position in international business. The counterpart in science is the Royal Swedish Academy of Sciences which is distinguished by its administration of the prestigious Nobel Prizes.

The Fraunhofer Society (FhG) is a major technology institute in Germany. It is currently comprised of 67 member research centers, laboratories and institutes.[90] It was established after WWII in 1949, with headquarters in Munich for the post-war construction of Germany. The Fraunhofer Society grew rapidly after the reunification of Germany in 1993. It now employs about 24,000 professionals, with R&D centers spread throughout the German Republic. Its technical fields cover energy, materials, microelectronics, optoelectronics, personal computers, computer software, automation, robotics, sensors, wireless, environmental protection, biomedical and health sciences, etc.

Organizationally, FhG is a non-government organization (NGO) with basic funding from the state and local governments but it receives about 70% of its budget through contract work from industry and government-sponsored projects. This is a very good indication of its close interaction with German industry and the university. The Fraunhofer Society is the counterpart of the Max Planck Institute, which focuses on basic science. The active role FhG has played in establishing Germany as an innovation country is well recognized.

In June 2011, the U.S. Council of Advisors on Science and Technology, in a report to President Obama, recommended policy measures to be taken for ensuring American leadership on advanced manufacturing,[91] pointing to FhG and ITRI as examples of innovation.

Industrial Technology Research Institute (ITRI) was chartered by the Republic of China Legislature in 1973 as a private entity whose mission was to advance the technological level of Taiwan's industries. It received basic funding from the government as a national institution but was supervised by a Board of Directors and operated as a private entity. ITRI employs about 6,000 professionals and has its headquarters in Hsinchu, with branch offices in many other cities.[92] Its budgets reached a 50/50 goal from public/private sectors in 1995. Its technology efforts covered the fields of electronics, personal computers, computer software, communications, wireless technology, advanced materials, optoelectronic devices, specialty chemicals, industrial automation, industrial robots, energy conservation, green energy, industrial pollution, industrial safety and bio-medical devices, all of which are relevant to the development of Taiwan. These efforts have been instrumental in transforming Taiwan from an agricultural economy to an industrial economy and to a knowledge economy in the last three decades.

It should be noted that both FhG and ITRI were selected as among the top 100 most innovative organizations worldwide by Thomson Reuters in 2013. The operation model of ITRI has been emulated in the Asia Pacific region. Some salient features of ITRI will be discussed in Chapter 17 of this book.

In summary, the technology institute is an integral and important part of the technological capacity of a nation. There are many institutes in the world that were recognized as important players in the innovation systems of nations between the second half of the 20th century and the present. The list is long and will include Agency for Science, Technology and Research (A*STAR) of Singapore, Chinese Academy of Sciences (CAS) of China, the Commonwealth Scientific and Industrial Research Organization (CSIRO) of Australia, the Japan Institute of Advanced Science and Technology (JAIST) of Japan, the Korea Institute of Science and Technology (KIST) of South Korea, the Netherlands Organization for Applied Scientific Research (TNO) of the Netherlands, and many others. The role each plays in their respective nation is an important subject that merits further research.

Part Three
Challenge and Change

By pursuing innovation and entrepreneurship, a professional will step outside of the academic ivory tower into society. He will encounter the challenges of people in the real world who are of diverse background and different interest. The challenges may appear to be overwhelming at times. How will he cope with this vast change of environment?

I was greatly inspired by the teachings of Laozi and Confucius in Chinese history.

Laozi has provided insightful observations on the relationship of man and nature. He taught people to learn from nature and abide by its rule. His teaching shed great light on the source of innovation and how the real and virtual attributes of things relate.

Confucius studied man and society. He taught people to interact with each other based on respect, reciprocity and trust. His teaching is the foundation of corporate governance and management science. As such it is fundamental to entrepreneurship.

I will discuss how these teachings can be used as guidance to help an entrepreneur when he has to make hard choices in the face of challenges.

Chapter 12

THE POWER OF REAL AND VIRTUAL: LAOZI

Learning Innovation from Laozi

For many years I have pondered on how to make the innovation process for new products more efficient, their value more versatile, their useful life longer and final disposal less harmful to the environment. This was prompted by the observation that a new product to be displayed in a showroom takes many years of development. Its usefulness is limited but its maintenance is costly and its disposal generally traumatic.

Take automobiles as an example. It takes three to five years to develop. Its range of speed is rather limited, its useful life usually no more than 20 years and its final disposal generally complicated and polluting. The electronic computer may be a little better. But it still takes two to three years to develop. Its use is more versatile but it needs frequent upgrades. And at the end, its disposal is complex, part of it may be toxic.

Take another example on a smaller physical scale: pharmaceuticals or fine chemicals. They take five to ten years to develop, their usage is very specific, they need reinforcements or replacements from time to time. And worse still, they may be found to be cancer causing or hazardous to life in ways that were never suspected 30 or 40 years later.

These three examples represent three categories of manufactured products found in our daily life. Let us look at another product (of nature) — the human being. It takes nine months to create, can be developed to sing songs, paint paintings, run like a horse, swim like a fish, operate machines, manage businesses and even run the country, with blunders or stumbles, notwithstanding. And, at the end of 60–80 years of useful life, ashes to ashes.

Man should take lessons from nature and abide by nature's rules. This is the theme of the teaching of Laozi (~600–530 BC). His famous work, the *Tao De Jin* (道德經) is one of the most precious treasures of Chinese culture.

Confucius, who was about 30 years younger, once paid him a visit and was asked by students to describe his impression. Confucius was quoted saying, "We know birds can fly and animals can run. Those that run will be stopped at barricades; those that fly will be trapped by nets. But the dragon will rise to the heaven with blasting wind that will leave you in awe. I saw Laozi today. He was like the dragon."

When I was a high school student I developed a great fascination with Laozi but stumbled at the very first paragraph of *Tao De Jin*, as many of my peers did. The passage defined the relationship of man with nature. It was so simple and mystical that it was beyond my comprehension. I gave up and left the book alone. Later on, I learned to take what it says as granted faith and continued with it out of curiosity. One day, when I was close to retirement as a professor of materials science, I happened to pick up the book and flipped the pages. On Chapter 11, on "Wu Yung" (無用), it reads:[93]

Thirty spokes cross at one hub,
Make the hub hollow to have the use of carriage;
Milling clay to make containers,
Make it empty for the capacity to hold;
Building houses with bricks and stones,
Cut doors and windows for the space to live.
Therefore,
The Yu defines the Li; the Wu provides the Yung.

"*Yu*" is "Have"; "*Wu*" is "Have Not". "*Li*" is "Use", "*Yung*" is "Usefulness". I was astounded to find that Laozi, whom I had thought a philosophical giant, spoke the language of an engineer. What a statement this is! "The Have defines the use; the Have Not provides the usefulness." In the broader sense, "*Yu*" is the real, the hard, the solid and the full; "*Wu*" is the virtual, the soft, the void and the empty. Laozi's conclusion was simple and striking. One needs the hard part to define the use of an object and the soft part to provide the object's usefulness. The hard and soft attributes are complementary to each other; they are integral parts of the whole.

Take a building as an example. The hard part, be it steel, aluminum or brick, or any combination thereof, defines the boundaries of the building in natural space. But it is the void, or empty space inside, that provides the utility of the building. It is this soft part that produces the revenue. The building can be a bank, office, hotel, simple residence or a combination thereof. No matter, it is the empty space inside the solid frame that enables human activities to take place and makes the building useful.

Look at a cargo container ship. Each container is simple hardware that defines the space available for goods to be stored and shipped from one harbor to the other. When the goods inside the container are unloaded, it will have the capacity to load other goods and generate more income. Thus the container shipper provides a useful service for product manufacturers to move the goods around the world. The shipper makes money on the empty space.

An example to illustrate the balance of hard and soft attributes of industrial products is the computer. All computers, regardless of their computing power and physical size, consist of hardware and software. The hardware includes the processor, monitor, keyboard, peripheral equipment and attachments that one can see. It also includes the microprocessor chip, printed circuit boards and other electronic components inside the computer case. But the value of the computer to the user is none of these. The value lies in the software that makes the computer compute, data be processed and results be presented.

The development and evolution of the computer has shown that over the last 50 years its physical size shrank significantly while its software expanded. 50 years ago it would take a 100-meter square room to house a Cray supercomputer; now the iPhone 6 with similar computing power can be held in one hand and put in the shirt pocket. The system software and apps have enabled the iPhone to do more for a normal user than the supercomputer could. In other words, while both the hardware and the software of the computer have improved over the years, the value of the computer lies more on the software, or what it can do for users. While the hardware has seemed to approach its limit, innovation with the software has no end in sight.

It is of interest to note that Joseph Needham in his classical treatise *Science and Civilization of China*, recognized Chapter 11 of Laozi for its discussion of *yu*, *wu* and *li*, *yung*. He places focus on the difficulty of translation and obscurity of the meaning of these fundamental concepts.[164] He did not, much to my surprise, seek a more in-depth application of the concepts in our daily life.

The Computer Industry as an Example

The focus on expanding computer software attributes is unmistakable. We can examine this industry's four segments: IC chips, computer assembly, software (system and application) and sales and distribution.[94] In the 1980s the computer industry was a vertically integrated one dominated by hardware manufacturers. For instance, IBM made the chips, assembled the computer, developed its software and sold the products. The users, individual or corporate, could only purchase IBM computers through the IBM sales department and its official representatives. Other giants in the computer industry such as DEC, Sperry-Univac and Wang Laboratory all operated in a similar fashion. The industry was in the hands of those who had control of the hardware.

Come the 1990s, the computer industry turned horizontal. There are many other players in the four segments. For IC manufacturing, apart from IBM, there is Intel, Motorola, newcomers such as TSMC, Infineon, Samsung, and others. For computer assembly, there are important players like Apple, Compaq, Hewlett-Packard, Dell, Acer, Asus and others operating on all continents of the world. In computer software, Microsoft and Sun Microsystems were recognized leaders followed by numerous players offering specific application software. On sale and distribution, computers are available in retail stores, superstores and various electronic commerce outlets. The hardware is just a part of the computer industry. Computer giants of earlier decades, such as DEC, Sperry and Wang-On that failed to recognize this trend and adjust to it gradually disappeared from the scene.

In the new century, globalization has advanced rapidly, making manufacturing of all computer hardware parts and components scatter all over the world. For example, Intel has established major facilities in Southeast Asia and Central America. The monitors, all liquid crystal display (LCD) based, were manufactured in China, Japan, Korea and Taiwan. All the major computer companies have set up manufacturing centers around the world. The job of a computer company has become setting specification, sourcing and qualifying suppliers, system integration, testing and packaging, and management of the global supply chain. These soft attributes of management determine the success and profitability of a computer company.

The last decade has seen the advent of cloud computing. While the hard attributes of the computer are important, the soft attributes take the dominant role.

Cloud Computing

Cloud computing brings about fundamental changes in the use of computers and computing. It is a business model that provides users with ready access to a shared pool of computing resources. The users, large or small organizations, have no need to establish their own "computer centers" with expensive infrastructure and maintenance. Instead, computing services are available at any time and in any place on a "plug and play" basis similar to our access to electricity from the power grid.

Thus infrastructure, platform and specific software, all on a service basis, will be available to the user through the cloud called IaaS, PaaS and SaaS, respectively. It can include data, data storage, microprocessors and devices, servers and networks, system software and application specific services.

Access is made possible through internet technology, with flexibility and elasticity via desktops, laptops, flat-pads, television, flat screen, mobile phones and other wireless devices. The services can be for personal, general, business and any combination of uses.

Hence cloud computing represents the sharing of computer and computing power on a global scale and is efficient and environmentally friendly. At the firm level, it reduces operation cost, improves efficiency and allows almost unlimited sourcing from suppliers and distribution to clients. It facilitates new innovations from lab to market and brings fruits to mankind. Therefore it will have enormous impacts on human lifestyles and how business is conducted.[95]

Think First

It is of interest to recall another thought of Laozi. In Chapter 40,[96] he said,

The Dao functions in cycles,
    It shows its role through the weak;
Everything in this world has its existence in the real,
    The real originates from the virtual.

"*Dao*" is "Way" or "Natural Rule." I think that an excellent illustration of the last two sentences is cloud computing.

In cloud computing, everything operates in the void, or the virtual, but its usefulness is accomplished at the user end, the real. It is a powerful tool to augment the effects and benefits of all technology innovations. The rapid access of cloud computing and its ability to manage and analyze vast amount of data, or big data, are all manifested in the virtual.

In the last decade, innovations and entrepreneurship based on cloud computing have been shown to play fundamental roles in our lives and bring about many social changes. Numerous examples can be found in the following areas: knowledge provision, entertainment, electronic commerce and social networks.

Knowledge provision is now a part of our daily routines. Every time a new phrase comes up in reading or conversation, one consults Google or Wikipedia or Baidu for its meaning and for background information. When one travels to a new place, one consults Google Maps or other similar apps. The latest automobile models provide information and directions to the nearest gas station, restaurant or hotel, at the driver's fingertips. All the data processing and computation are done in the virtual but the impacts are manifested in the reality of daily life. It is little wonder that companies such as Amazon, Microsoft, Facebook, Google, Baidu, Tencent, Twitter and Alibaba have dedicated a great deal of effort to the virtual community.

The Future Virtual World

Another example is intelligent robotics. Traditional robots are designed to do simple repetitive work, work beyond human capability, or work in unfriendly or stressful environments. However, modern robotics is concerned with giving robots cognitive capabilities such as anticipation, planning, attention and perception processing, in addition to complex motor coordination. In other words, the robots will be equipped with intelligence and able to manage more complex problems in the work environment. In fact, all previous work experience will be recorded, stored and classified inside the robots. This will give the robots a new level of intelligence.[97] Thus, the robot will be more intelligent as it grows in use. Remember the saying, the older the wiser? These will obviously require a great deal of design, analysis and reiterations in the virtual world. Thus the working robot exists in the real but its power, the intelligence, comes from the virtual.

In this connection, I should admit that there is an unsettling feeling with intelligent robotics. Will it replace humans in the long run? It is quite conceivable that one day the intelligence of a robot will approach or even exceed that of man. And, when this happens, the robot will be wiser and younger, or most definitively, wiser and physically stronger than man. How will man be safe from the robots? Will man have to escape to the cave for safety, where there is no electricity from the grid or the cloud to energize the robot? We will have to search for what Laozi might have said in this scenario.

Competitiveness of Innovation

The power of the virtual is not limited to computer science alone. The ability to think virtual is at the heart of competitiveness of innovations in any field. One needs to place oneself in the virtual world, adapt to the new environment and envisage scenarios on how to play a useful role. Thus the competitiveness of innovation will comprise the following issues.

- Need: the niche advantage
- State-of-the-art hardware
- State-of-the-art software
- Simplified processes
- Upgradable values-in-use
  - versatility
  - scalability
- Miniaturization
- Virtualization
- Quality in delivery and services
- Zero Carbon Footprint
- Use cycle management

Take the case of chemical engineering, as shown in Figure 12. To help improve the quality of life, one would frequently think about the hard sciences, such as formulating a new material or material system and developing it into a new product through some novel processes. Following Laozi's suggestion, one should first closely review the environment, analyze the conditions and scenarios, and devise options to realize or augment the need. There may be computer models that can be used to demonstrate the concept and simulate actual applications. In this process, classical chemical engineering training in kinetics, thermodynamics and transport phenomenon can be called in to help formulate desirable solutions. Live experiments can be performed afterward.

Robert Langer, one of the most cited chemical engineers in history, is Director of the largest biomedical engineering lab in MIT. He has done outstanding research in drug delivery, which has led to many new entrepreneurial companies.[98] His research was based, to a great extent, on the ability to visualize the environment of the drugs that would be captured and stored inside the body and, at defined intervals, released in response to external signals. The accomplishment is real and has its origin in the virtual.

The power of the virtual is not only in technology and business but also in politics. The 2016 presidential election of the U.S. demonstrated that Donald Trump's victory (based on his own assessment) was helped by social media: Twitter and Facebook.[99] In direct communication through these media, individual voters had the satisfaction of being informed and consulted by the candidate and became associated with his position. Thus the soft attribute of the computer was explored and exploited to its extreme. The only thing I would add is this: the power of information flow to the people, like the power of water flow to a ship, can float as well as sink. In the end, it is the respect and trust that determines the fate of politicians in history.

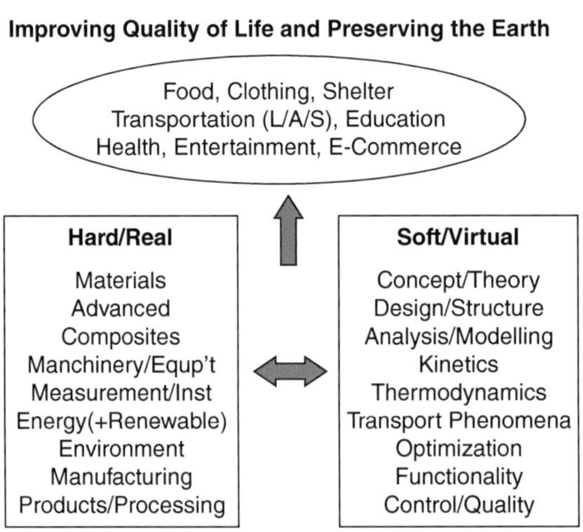

Figure 12. The Hard and Soft Science of Chemical Engineering

Chapter 13

FROM SELF TO SOCIETY: CONFUCIUS

Learning Entrepreneurship from Confucius

In a broad sense, entrepreneurship is the pursuit of an innovative idea to create value for the public while knowing that the process involves many serious risks. From this perspective, Confucius (551–479 BC) was an avid entrepreneur; he had many ideas he believed would be beneficial to individuals and the society at large. He made great efforts to find opportunities for their implementation, to no avail. Recognizing that his vision might be too advanced for his time, he travelled around various states and took in students to nurture them as future leaders. He hoped that, perhaps, one or more of the students would be more fortunate than he was and opportunities would arise for the society to implement his political ideas. Confucius has long been recognized as a role model of educators in China. In my view, he has made much more important contributions to mankind as a pioneer of democracy and human rights, and advocate of social harmony.

Soft Power

As a teacher, Confucius was friendly, honest, humble and personable. He entertained questions from students and encouraged discussion and debates. One such session went as follows:[100]

> *Tze-Kung asked what was needed for the governance of a nation. Confucius said, "Adequate food, adequate arms and the trust of the people."*

Tze-Kung asked, "What if I am forced to drop one, which among the three should go first?" Confucius said, "Arms."

Tze-Kung continued, "What if I am forced to drop another, which one should then go?" To this, Confucius said, "Food. Death is inevitable but the State cannot stand without trust."

Confucius' conclusion was astounding. Food, which represented economic power, and arms, which represented military strength, were the hard power supporting pillars of a nation. Their importance was obvious to governments. Trust of the people constituted a "soft power." It is certainly desirable to have it. But if it were not there, one would think, nobody would miss it. How could it compare in importance with the other two?

Confucius saw it differently. His view was that trust of the people was the base line that governments should never betray. No government could do anything without trust of the governed. Trust is the foundation of national governance.

Any student of history will recognize that governments come and go and nations rise and fall when famines, wars or disasters strike. But the root cause of a nation's demise is not famine or war or disaster. It is the loss of trust. In difficult times, people will tighten their belts and fight with bare hands to defend their country if they trust their king. But when the trust is gone, nothing can save even such figures as the Emperor of the powerful Chin Dynasty, the King of Prussia or President of South Korea.

In placing the priority on soft power rather than hard power for national governance, Confucius made his mark in global political literature.

If one looks into it more deeply one will find that trust is intrinsic to even hard power. In economic terms, trust reflects a nation's credit and creditability. It relates directly to the cost of money and the ability of a government to build and grow. In military terms, trust generates faith, mission and responsibility.[101] It is at the very basis of command and discipline. Therefore trust is the deepest foundation of hard power.

Trust in Entrepreneurship

Placing priority on trust is also important to entrepreneurship and business organizations. Numerous examples can be cited in the business world. Chapter 20 of this book tells a little-known story of the world's bestselling carbon fiber bicycle company. In the spring of 1990, the first batch of products was launched for a trial, but it was soon discovered that the joints of some bicycles might loosen with use. On hearing this first report of a possible defect, K. B. Liu, the company president, made a swift decision to pull all the trial bicycles off the floor. He ordered that the problem be resolved before any product was back on the market. He demonstrated how much he valued customers' trust in his brand name. It was not just an individual case of liability. Product safety was also an issue of social responsibility.

In his effort to resolve the technical uncertainties, he improved the manufacturing process and quality assurance practice. As a result, the company was able to gather proprietary information on the technology of carbon fiber bicycles. This gave it an unmatched advantage on the market. The small company, GIANT Bicycles, has ultimately become the number one carbon fiber bicycle manufacturer in the global market.

Chapter 22 also describes the story of Ardentec, a firm doing business in integrated circuits (IC) testing, and how it has carved out a strategic position in the semiconductor manufacture value chain. This company's business model rests precisely on the trust of high-tech companies at both ends of the value chain. The supplier and the recipient of ultra-large-scale integrated circuits (ULSI) rely on the third-party IC tester to be an honest broker in their transactions. This entrepreneurial company strives for excellent quality and absolute confidentiality, which are components of trust.

There are unbelievable stories on the flip side. In late 2013 observers of the business world were amazed to learn that a defective ignition switch had existed for seven to eight years in nearly 25 million automobiles manufactured by General Motors (GM).[102] It was reported that the defect was a minor one (with a replacement cost of under US$1) but it could cause drivers to lose control of their vehicles and lead to accidents and unimaginable misery.

Reportedly, over 200 accidents and 70 deaths had resulted in the course of the years. GM was fined nearly 1 billion dollars by the U.S. government. Numerous lawsuits brought by families of the casualties are still pending.

The incident has been a severe blow to the trust and reputation of General Motors, which was once the world's number one automobile manufacturer. Its effects on GM sales and the company itself will probably be a subject of further research.

Not to be outdone, Volkswagen AG, which prides itself on automotive engineering, was found in 2015 to have altered computer software that monitored exhaust fumes and fuel efficiency.[103] As a result, its diesel engine cars could emit nearly 10 times more nitrous oxide than the law allowed. What made the matter unbelievable was that the practice was apparently known to top management for some time. In the end, the president of the company was forced to resign. In January 2017, six Volkswagen executives were indicted by the U.S. District Attorney's Office for criminal offenses in similar cases in the U.S. And, yet, Volkswagen's problem will not end with this because the litigation in Europe has just begun.[104] The damage to the trust of this popular and proud company is hard to estimate.

Every time I read the Confucius–Tze-Kung conversation, I am deeply touched by Confucius' conclusion on the supremacy of trust as well as the teacher-student relationship. This kind of direct and honest exchange signified the mutual respect and pursuit of truth that existed between professor and student.

Man and Society

Confucius' theory on organizations and management placed responsibility squarely on the individuals in leadership. He stressed the importance of self-discipline, filial respect, family honor, love and forgiveness, dedication to jobs and governance, and loyalty to nation. He believed leaders should lead by practice, motivate people according to their individual needs, communicate the vision and plan ahead.

Confucius' respect for people was exemplary. He treated every question he received from students seriously and answered the question specifically tailored to the student's background and need. Many questions regarding the issues listed above were raised by students. Filial respect was a common topic: I have counted no less than 10 answers on records to 10 students who asked about it.

Regarding study and discipline, Confucius outlined a process necessary for preparing oneself for leadership in public service. The process starts at the level of the individual, and then progresses to family, country and world.

With respect to the individual, he placed the emphasis on research for new knowledge. He said an individual needed to engage in research in order to gain knowledge. Knowledge would make the individual humble which would help the individual focus on the priorities in life.[105]

I believe what he said is pertinent to technology innovation and entrepreneurship. Entrepreneurship does not come from ignorance or arrogance. It takes knowledge to acquire the understanding of self, assessment of strength, focus on priority, as well as to plan the course forward.

Strive for Excellence

Confucius set the goal of study as the search for excellence and provided guidance for implementation.

> *Study broadly, inquire carefully, think judiciously, distinguish clearly and execute resolutely.*

I believe this may have been the earliest publication on the principle for the search of excellence.[106]

Confucius' guidance was translated into practice in modern quality engineering and management methodologies. Notable amongst these is the manufacturing systems of the Japanese Toyota Automobile Manufacturing Corporation popular in the post-WWII period. It was based on a learning process known as LAMDA: Look, Ask, Model, Discuss and Act.

This has a striking similarity to the Confucian principle proposed some 2,500 years ago. Other popular quality and productivity management services such as Lean Thinking[107] also found their foundation there. Understanding the principle of quality engineering practices will go a long way in supporting entrepreneurship and corporate management.

A Model of Education

A major part of Confucius' life was dedicated to education. He took it as his mission to educate the common people to become scholars who would play leadership roles in the society and country. The words *chun zi* and *xiao ren* (君子與小人) appeared frequently in his vocabulary.

These two words are very complex and intricate to translate to English fully. Accomplished scholars in both Chinese and English literature like Joseph Needham had encountered the difficulty. He said, *chun zi* "is one of those words which I have come to the conclusion better not translated." In its character, *chun zi* refers to someone who may be "well born, a scholar, an official, a soldier, a martyr." *Xiao ren* "is the opposite; a man not only of low social station, but also meanness, boorishness, etc."[164]

I would translate the words to "scholar" and "commoner." This translation is obviously imperfect; but I feel it can convey the salient features that Confucius expressed.

Some would argue that Confucius tended to stereotype people into classes by these two words. In my view, this is not true. Confucius simply pointed out that people are different. And people have different needs and motivations. But even those who are not born perfect in character could improve themselves via education. Leaders of the country or chief executives of companies needed to recognize that people are different, and that the same person could act differently with a change in needs and wants as the circumstance changes. It is the job of a good leader to understand the person and address the changing circumstance.

Confucius was quoted as saying, "Scholars can be reasoned with ideal and principle while others (commoners) with cost and benefit."[108] He meant that there are a range of needs and wants in people. Therefore in an organization the leader should adopt different approaches to address different needs when working with different people.

For example: everybody knows that drivers should not overspeed because it is dangerous to themselves and to others. Everyone should behave prudently and carefully. This is the ideal and principle. But governments set up speed checks with punishments for those who violate the speed limit. This is based on the cost-benefit perspective and is more effective in achieving safety on the roads.

On the positive side, students should know that it is their responsibility to study hard and gain scholastic achievements. This is ideal and principle. But schools usually give honors and prizes to those students who have done exceptionally well and met a certain standard. This does not mean that good students should decline honors or prizes. Getting an award or prize will in no way reduce the student.

In any case, one should recognize that people are complex and individuals need different motivations at different times. An entrepreneur needs to prepare a business plan for investors. This should include the vision and objective of the venture, but also its costs and benefits. He needs a business plan with different versions for different audiences. An angel investor would be happy to share the vision and objective and will generally take the entrepreneur's word as is. Venture capitalists and bankers need to see a cost-benefit analysis and often, expected returns on investment and an exit plan. It is not a matter of good or evil, but a matter of needs and motivation. This is all human nature!

Management actions for satisfying the different needs and wants of people in organizations are among the main themes of modern management sciences.[5] The issue of different personalities and motivations was deftly addressed by Abraham Maslow.[109] Maslow showed that there is a hierarchy of needs in people and that the management should take action to help people fulfill those needs. This is in line with the thoughts of Confucius on motivation.

While I served the ITRI in Taiwan as president in the mid-1980s, I instituted a series of management improvement programs based on Maslow's hierarchy of needs[62] in recognition of Confucius' teachings on human nature. I am convinced that ITRI's later performance was greatly helped by this corporate culture.

For entrepreneurs who have proficiency in the Chinese language, a reading of Confucius and his third generation student, Mencius (孟子), would be highly enlightening.

Value and Ethics

All entrepreneurs want to make their ventures successful; but this is a very difficult challenge. In scientific research, one deals with the object of study with certain methodology and experimentation under conditions one has full control. In innovation, one needs to go outside of the laboratory to focus on the changing demands of the market. In entrepreneurship, one needs to be in the market and accomplish business objectives through the actions of many people. The challenge is exponential.

Finally, let me quote another exchange between Confucius with Tze-Kung.[110]

> *Tze-Kung asked, "Is there one word that can guide one for the entire lifetime?"*
>
> *Confucius said, "That would be Shu* (恕)! *Do not do unto others what you do not want others do unto you."*

This is amazing, a discussion of what in modern time is referred to as universal value or eternal value. Not fanciful words like democracy, freedom, rule of law, human rights and social harmony. These are normally used in the Western world. Or loyalty, patriotism, unity, economic growth, etc. These are generally respected in the Eastern world. But, according to Confucius, it was "*Do not do unto others what you do not want others do unto you.*" How startling! Needham had translated *shu* as forgiveness.[166] It was very close although I would rather not translate it!

To think about it, "*shu*" involves more than respect or forgiveness for the individual. It is not just the principle, it is the practice. So while one wants to protect one's home, family, job and aspirations, help your colleagues, employees and other stakeholders to do the same. This is not just management science, this is religious ideal. So, have respect and trust people. This will not only help one to be a successful entrepreneur but also a good person.

Chapter 14
MANAGING CHANGE

The Nature of Change

The Chinese poet Xu Dong-bo of the Sung Dynasty, while sailing down the Yangtze River, composed an emotional essay[111] paying tribute to the old battlefield *Che Bi* (赤壁), in which he talked about a paradox of life: change.

> *Have you not noticed the water and the moon? The water flows but never disappears; the moon ebbs but never diminishes. Viewed from the perspective of change, the universe has never stood still even in the wink of an eye. Viewed from the perspective of constancy, nature and man will exist forever.*

Change is the subject of *Yi Jin* (易經, literally, *The Book on Change*) which is probably the oldest known classic of China. As a documentation of the laws of nature and their effects on man, it was reportedly written by three authors nearly a thousand years apart, the latest being Confucius. Many ancient scholars also studied *Yi Jin* with varying degrees of interest in application areas such as astrology, medicine, weather, Feng shui and event forecasting. Laozi interpreted many facets of nature and speculated on their relevance to human behavior.

With regard to innovation and entrepreneurship, the most important lessons from the sages, I think, are as follows.

> **One: change is inevitable. Two: change is cyclical. Three: change favors the weak.**

Change is inevitable because the natural celestial system is dynamic and functions under complex rules. To name just one: the earth revolves around the sun and the moon around the earth at the same time, in different orbits. The earth, through the force of gravitation, brings air currents, bodies of water and all the goodies inside as it moves around. This sets the stage for the climate and the dynamics of the earth's eco-system.

On the individual human level: metabolism goes on and living components get oxidized as the body ages. And most complex of all, inside the human body, physiological and psychological changes take place quite mysteriously and affect people's behavior. Thus changes may be small and gradual, but are most certainly continuous and cumulative. In the end, the accumulated changes become apparent and the result is irreversible, a surprise to no one.

Change and Entrepreneurship

A smart entrepreneur is one who can discern and detect the smallest changes in the market or organization that may affect the value of his product or service. He will make allowances for, or adapt to, change. This is not at all an easy task. For scientific research, one only needs to work studiously and assess any changes in the lab. For technology or innovation, the changes are more complex because the focus is outside of the laboratory. For entrepreneurship, the job is formidable since the battlefield is full of changes and uncertainties involving customers and competitors in national and global markets.

Change is cyclical because the universe is in a dynamic balance. The sun rises in the east and sets in the west; night follows day, and the pattern repeats itself. In agriculture, a year has four seasons. The spring for incubating and growth, the summer for blooming and producing, the autumn for harvesting and reaping, and the winter for storage and recuperation. Then the cycles of the year recur.

Industrial businesses move up and down in similar fashion but with their own cycles. A technology start-up incubates, grows, matures and deteriorates. It is a difficult challenge for the entrepreneur because he will have to manage sufficient resources to sustain the incubation and growth cycles. He will have to convince investors that while the revenue is small at the start, it will continue and become larger and larger as the business matures.

More often than not, start-ups cannot sustain themselves long enough for the revenue stream to become real, or real enough to entice investors to provide additional funding. This is the great challenge of entrepreneurship. Thus a good entrepreneur is one who can structure the company with a portfolio comprised of a group of related products and services. They will be at different technology phases, some in their early stages, others in the maturing stages. Thus, with overlapping cycles and revenue streams, the overall business will be in growing mode, despite the ups and downs of the components.

Change as an Opportunity

That change favors the weak (see Chapter 12) is not an easy concept. One has to reach deeper to grasp its significance. Take for example, a pendulum. It appears to be weakest and nearing standstill at its lowest point. But precisely at this point it reaches its maximum acceleration enabling it to swing to the other side. When, at the far side, the speed becomes zero and the pendulum appears to be at a standstill, the potential energy makes it reverse direction and repeat the process. Of course the acceleration and potential energy are things the eyes cannot see. They are the apparent weak components of the motion but they drive the process and lead to further changes. In Laozi's term, the weak becomes the dominant trait of life. It is nature's function to nurture and grow the weak and the young in the process of life.

The advent of the internet and wireless technologies in the last decade has undoubtedly speeded up social changes at a greater pace and scope than ever before. In the global village, the backyard of one country is the front porch of another. Pollutants and terrorist agents will not stop at the borders; they travel around the world with exceptional frequency and ferocity. Thus change in one place creates ripple effects in others, which brings unexpected opportunity in yet other places.

The rise of social media such as Facebook, LinkedIn, WeChat and Twitter in the 21st century further increase the changes. They have had tremendous impact not only on the economy but also in society and politics. They appear to be equalizers in the digital divide and great enhancers for the general public. People would respond to quick jolts of emotion rather than reason with calm logic. Do rapid changes like these cause more harm than good in society? The jury is still out. It appears that these innovations are double-edged swords. We will need to pay more attention to their implications before it is too late.

That change favors the weak has also offered newer and greater opportunities to novices in society, especially in populist politics. The unusual success of Rodrigo Duterte of the Philippines and Donald Trump of the U.S. are good examples. At the beginning of the race, they appeared to be underdogs. But their simple appeals to emotion and change, fueled by bluntness and plain words, identified them with the yearning for change on the part of the masses. Thus the apparent weaker contender was propelled to the top, much to the surprise of expert observers and pundits. Actually, Barack Obama was also viewed as the weaker candidate in the early part of the 2008 U.S. presidential race. The same force favored the weak and he was elected.

In reality, the effect of change should not come as a surprise. The changes started as minor and gradual but are continuous and persistent. Eventually they created such turbulence as to change the course of the world. But the elites, the strong and powerful, were generally too complacent or insensitive to discern what was happening and thus got the short end of the stick.

Crisis Brings Opportunity: Examples in People

The clearest statement on the interdependence of change — good and bad, success and failure — can be found in Laozi.[112] In Chapter 58, he said,

> *Misfortune breeds fortune. Fortune hides misfortune.*

History is loaded with examples. Perhaps one of the best-known personal stories concerns Franklin D. Roosevelt (1882–1945). He was born to a prominent family in New York and graduated from Harvard University. He entered politics at age 28, serving in the New York State Senate and then as Assistant Secretary of the Navy. In 1920, he was chosen by the Democratic Party to be the running mate of the presidential candidate James Cox, only to lose to the Republican Harding-Coolidge team. Nevertheless, he gained national fame as a rising star. But on a pleasant family trip in 1921, he was stricken with debilitating polio. As a result, he lost the use of his legs and was restricted to a wheelchair. His political career appeared to be plunging to great jeopardy. What a blow to an ambitious young man this was.

But Franklin Roosevelt was not discouraged by the misfortune. He worked diligently to reconcile himself with his new reality and mastered his heart and mind sufficiently to move on. In 1928, he ran successfully for Governor of New York and worked unwaveringly to help combat the Great Depression that was confronting the country at the time.

In 1932, Roosevelt ran as the Democratic presidential candidate and defeated the incumbent President Herbert Hoover in a landslide victory. During his first 100 days of presidency, Roosevelt spearheaded major legislation and instituted the New Deal to revitalize the country. The U.S. economy improved rapidly and he won the re-election in 1936. A series of programs he initiated to guide socio-economic development, such as the Securities and Exchange Commission (SEC), Federal Deposit Insurance Commission (FDIC) and Social Security, are still the sustaining fabric of the U.S. society today. When WWII broke out in 1938, Roosevelt decided the United States would be the "Arsenal of Democracy" and played a leadership role in fighting the Axis Powers of Imperial Japan-Nazi Germany-Fascist Italy. He died in 1945.

Franklin Roosevelt's 24 years of public service after polio struck marked a high point in U.S. history. He is considered by most historians as one of the top five U.S. presidents. This supports Laozi's principle of misfortune breeds fortune.

There are other stories of people who suffered misfortune but fought it with positive attitudes and resolute actions. For example, Helen Keller, who became blind and deaf at 19 months old, fought her whole life (aged 88 at death) to become a great writer and advocator of socialism, women's suffrage and human rights. Ludwig van Beethoven who, despite hearing loss and health deterioration mid-career, continued to compose sonatas, concertos, quartets and symphonies including the 9th which is considered the crown jewel of symphonies.

In China, during the Western Han Dynasty, an outstanding author and scholar Sima Qian[113] suffered the gravest insult and bodily harm inflicted by the injustice of the Emperor. But he decided to endure the misfortune to continue his mission of writing the book *The Records of the Grand Historian*: *Kings, Ministers and Scholars* (史記). The book became a landmark document of Chinese history. The wisdom that it preserves and transmits is monumental in Chinese culture.

Crisis Brings Opportunity: Examples in Business

A classic example of misfortune breeding fortune in corporate management is the case of the Tylenol murders in 1982. It was found at the time that several individuals died in the metropolitan Chicago area after ingestion of Extra Strength Tylenol, a widely used over-the-counter pain reliever and fever reducer. Apparently, the tablets were deliberately contaminated with cyanide. Its manufacturer, the Johnson and Johnson Company, immediately pulled all 31 million bottles of tablets off the shelves of all retailers in the U.S. It was the first major product recall in American history. The company quickly designed a tamper-resistant gelatin covered capsule called Tylenol Gelcaps and sold them in triple sealed safety containers. Ten weeks after the recall, the new Tylenol was back on the shelf. The incident had cost the company US$100 million but Tylenol regained 100% of market share. It remained the top seller in the painkiller market in North America. Many surveys have shown that Tylenol is considered to be the drug that American doctors and hospitals trust the most. Thus the temporary misfortune to Johnson and Johnson helped solidify the company's leadership position in the drug industry and made it a role model for corporate responsibility.

Fortune Hides Misfortune: Examples in Business

We can now turn to successful companies indulging on good times. Like people, sitting on their laurels makes companies complacent and arrogant, which most certainly leads to failure.

Fifty years ago General Motors took pride as the number one auto manufacturer in the world. It produced 12 brands to suit the taste of every car lover. A high ranking U.S. government official once commented, "What's good for GM is good for the U.S." Volkswagen, a popular car producer in Germany, gained worldwide reputation for its engineering design and customer-oriented craftsmanship. Volkswagen owners are among the most loyal customers in the industry. How are these companies performing now?

General Motors (GM) suffered bankruptcy around the year 2000 and slowly recovered with government assistance. While it has been difficult to return to its past glory, GM has been doing quite well since. Then, in 2014, it was suddenly discovered that a large number of the cars it had made in the previous seven years contained a small defective ignition switch, which could trigger the electric system to lose power and cause fatal accidents. More than 29 million GM cars were recalled for repair. The company was fined nearly a billion US dollars by the government and underwent a great deal of litigation for its culpability. The estimated cost might well be over US$10 billion.[102] The question to ask is how a mistake that was widespread for so long had escaped the scrutiny of the management systems of GM.

Volkswagen, on the other hand, was found in 2015 to have used a software system that faked official exhaust tests and caused its popular diesel cars to let out 10 times more toxic NOx while running on the roads. It was further revealed that the top management actually knew and acquiesced in the practice. Thus the CEO was forced out, the company fined by many countries and its sales dwindled. The adverse impact will not be limited to Volkswagen alone but to the diesel engine related industries in Europe and elsewhere. Again, the fiasco was a corporate disgrace that bordered on criminality.[103] How could a company that prides itself on technological superiority end up betraying its own integrity and public trust?

A major scandal of corporate trust came to light recently in the Japanese business community. News in early October, 2017 showed that Kobe Steel, the leading metal manufacturer in Japan with 112 years of history was found to have falsified material data on its products for many years. The company under the brand Kobelco supplied about half of the global market of metal wires used in valve springs of automobile engines. Other usages of its products included mechanical components in aircraft hulls, nuclear reactors, air conditioners and microchips, etc. The user list, with more than 500 names, included Japanese Railway, Toyota, Nissan, Mitsubishi Heavy Industries, General Motors, Ford, Boeing and others. Thus the damages of such travesty may be serious, widespread and take a very long time for reparation. As an example, the Japanese Railway Company announced that 310 components of their bullet train were discovered to contain sub-standard parts from Kobe Steel. After the news broke, the share price of the company fell about 30% in 10 days.[183,184]

Hiroya Kawasaki, Chief Executive of Kobe Steel, admitted, "As yet, I have no clear reason why the data fabrication has happened."[185] He joined Kobe Steel in the 1980s and led the company since 2013. He was awarded the highest honour from the Japan Institute of Iron and Steel in early 2017. It was pathetic to see an icon of the Japanese industry let bare his negligence of duty, to say the least. This reflects the crisis of governance and trust confronting the Japanese industry.

Laozi did not advocate the acceptance of misfortune or fortune blindly as it came, he urged people and organizations to take good care in any circumstance and always behave positively and seriously. Life is a continuing process and the current situation, whether good or bad, will change sooner or later. Therefore one should be prepared for the next change to come and try diligently to make the change in one's favor.

Hence, in good times, one should act with humility and integrity, strive for excellence, and be prepared for future downturns with options. In bad times, one should remain assertive, creative, vigilant and courageous. This is the essence of the management of change.

Chapter 15

CULTURE OF ENTREPRENEURSHIP

I have personally seen hundreds of entrepreneurs in the last 25 years. Some have become successful, achieving results well beyond original expectations. But more have failed and many have changed career paths altogether. I also observed that, with the start-ups that are successful, most have changed team members and its championship along the way. Only a small portion have their original founders still with the company. So it was of great curiosity to me to assess the challenge and choices leading to the success of entrepreneurship.

Technology, team, timing, business plan, financial backing, all are important ingredients. But these are necessary conditions, not sufficient conditions. What is missing is the culture of entrepreneurship. Unfortunately this was seldom taught at any university so young people had to learn it the hard way. When they finally got it, it would seem only natural.

How does one define culture? In this context, it is similar to defining value. What does one value the most in life? For the entrepreneur, what are the values that really matter? More importantly, if the values are challenged by adverse reality, what choices should one make?

Cultural Change in the U.S.

To explore this question, let us go back a little to examine the recent history of the U.S. There are sobering lessons to be learned. In the first two or three decades of the 20th century, the U.S. created many important innovations for mankind. Examples include: Thomas Edison with electric light; Henry

Ford with automobiles; Alexander Bell with telephones; Wilbur and Orville Wright with the flying machine. These were defining innovations for the world.

Coming out of economic hard times of the Great Depression and two world wars, American people, by and large, were diligent, responsible, thrifty, trustworthy, empathetic and grateful for the opportunities that the new technologies provided them. The new knowledge and applications have motivated and driven the economy to grow extremely rapidly. The society was full of hope and goodwill, notwithstanding minor gaps and problems.

After WWII the U.S., through the support of a responsive government and academic institutions, was on its way to becoming leader of the world. The environment fueled the further growth of innovations in the following decades: semiconductors, computers, personal computers, internet, mobile communication, biotechnology, advanced materials, cloud computing, artificial intelligence, etc. It was the capacity for innovation and entrepreneurial culture that made the U.S. the strongest economy in the world.

By the end of the 20th century, changes in social culture have become apparent. To mention just a few:

- Play now, pay later
- Look at just the bottom line of the balance sheet, not the process
- Winner takes all — winning is everything
- No free lunch, but will take a chance for this one
- Throw-away society, things are abundant and cheap
- Push button for solution, no brain is needed
- See no evil, hear no evil — allowing gray areas of morality to stand
- Analysis paralysis — talks all the time, but takes no action

People are gradually getting sleazy, sloppy, sluggish and shortsighted. This is a culture entirely antagonistic to the values of innovation and entrepreneurship seen at the beginning of the century.

Thus the U.S. market and business have gradually become displaced or replaced by competitors from Europe, Japan, China, India and many newly developing countries whose people work harder and lead leaner lifestyles. The problems were compounded by inept governments, who plunged the country into war in the Middle East and by the wolves of Wall Street, whose actions led to financial bubbles and created a tsunami that swept through the world. Terrorist attacks and social injustice have become increasingly rampant.

Joseph E. Stiglitz, a Nobel laureate in economics at Columbia University, commented, "In the first three years of the so-called recovery after the 2008 financial crisis, 91 percent of the gains went to the top 1 percent of the earners."[114] No wonder the society has become greatly polarized and suspicious — which has led to further disarray in the 2016 presidential election.

What Constitutes Entrepreneurship?

The above three chapters of Part Four have discussed the philosophy of innovation and entrepreneurship from the perspectives of Chinese culture, especially that of Laozi and Confucius. It can be said that the culture of entrepreneurship can be understood from several standpoints.

- **People**

  People are the source of creativity and action. As employees, they need to be respected and motivated. As customers, they should be respected and their trust sought. Thus, any commitment made to the employee and customer needs to be fulfilled. The Confucian saying of "*Shu*" — do not do unto others what you do not want others do unto you — is extremely enlightening. This is a value that can be applied to people, organizations and countries. It is the source of harmony. It is the fundamental soil for entrepreneurship.

  Life examples can be found in all the cases discussed in Part Five of this book. Of special interest is the case of McDonald's Hong Kong. The founder said the focus of the business was not any food product; it was the people. People were the beef of his business. He would provide to the customers what they expected: value, quality, courtesy, consistency, cleanliness and promptness. He put a great deal of effort into training his employees to develop self-repect and to see themselves as crew members aboard a ship. He wanted the employees to take initiative in greeting and treating the customers as they wanted to be treated.

  In the case of Delta Electronics, the founder Bruce Cheng described respect for people as the key strategy of globalization. He said the employee in a foreign land needs to feel not just as an employee but also as a partner. He analyzed the different cultures and lifestyles of the local people in different countries. He adjusted his company's actions to address the needs of different localities.

- **Quality**

Quality is the result of striving for excellence. While perfection is the final goal, achieving the goal might take effort in phases of continued improvement. The Confucius methodology: study broadly, ask carefully, think cautiously, distinguish clearly, implement resolutely, can be executed in almost any company providing a product or service.

Perfection in quality is the most common denominator of successful products. The case of GIANT bicycles is a good illustration. Superior quality is the centerpiece of their brand. At the first sign of possible defects in the prototype the founder recalled all the products in the distribution channel. He wanted a clear resolution of the problem before the product went back on the market. In so doing, he also established unwittingly a proprietary processing technology that gave him an unmatched advantage that outpaced competitors.

Delta Electronics' command of the electronic parts and components market also rested on its insistence on superior quality. It continually tried to adopt a higher quality level before the market called for it. This provided the advantage that outpaced competitors.

- **Creating Niche**

Advantage in a certain niche area: performance, price, delivery, etc. will give a start-up a clear advantage in technology and the market. This is usually the result of having core strength in product manufacturing and delivery. Delta Electronics was able to excel as a lead supplier of parts and components to all television manufacturers in the 1970s/1980s. Then, for the emerging computer age, it foresaw the need for energy efficiency. It created a pivotal position for itself as the supplier of energy-efficient components for all PC manufacturers in the early 1990s.

The niche can also be something created in the existing value chain. Seeing the paradigm change in the electronics business, Ardentec fostered fab testing as a business separate from conventional manufacturing to make its client manufacturers more cost effective. This makes Ardentec a common virtual lab to the suppliers and end users of IC. This is a pivotal position to be in.

Likewise, Tencent was able to create a platform so that users could readily communicate with each other. This provided them with a unique position between the normal service provider and content provider. Alibaba.com's success was attributed to the ability to create different business models for SMEs and position itself in a niche area of the e-commerce value chain.

- **Adjusting Strategy**

The key elements in strategy are business model and timing. Business model relates to how products and services are packaged and delivered to the customers. And how the various resources of people, capital and technology are organized to support the business processes. More often than not, a venture will experience the unpleasant fact of the business model not playing out as originally designed. This is time to study the challenge objectively, however painful, find the cause and start anew.

Timing is of the essence. When encountering difficulties or challenges, it is easy for people to discuss and analyze and wait for the tide to change. It is often said that in the restaurant business, the key words are location, location, location. In technology entrepreneurship, the key words would be timing, timing, timing. Agility, taking swift timely action is extremely important.

In Part Five, one will find many examples of changing strategy to adjust to the marketplace or organization. The Epistar act of making changes in its corporate structure for wider access to the user market was a key step in its successful transformation. Tencent's response to the ICQ legal challenge and its agility in responding to users' needs for better service performance are exemplary.

- **Teamwork**

No one can do the job alone. A team with complementary skillsets and shared values is necessary for any start-up. The early executive team members of the Alibaba Group possessed talents in different fields: technology, finance, operation, market, and they came from different countries, which set the stage for its global presence. The division of responsibility among the five co-founders of Tencent highlighted the strength of teams. This was among the major factors responsible for their success.

- **Trust**

  Trust is the foundation of soft power for individuals, corporations and nations. The importance of trust from both individual and corporate customers is highlighted in all the cases shown. The different actions that GIANT Bicycle, Delta Electronics and Ardentec took to build up trust with their customers should be enlightening. This amounts to social responsibility and is the source of corporate sustainability.

  Many start-ups failed because of the erosion of trust between the leader and the team members, or, between the management team and the owner groups. A distinctive example of the former can be found in Silicon Valley during its initial phase. The Shockley Semiconductor Laboratories fell flat despite its technological superiority and capital support. Later on the loss of trust between the Southern California management team and the New York headquarters caused the fragmentation of Fairchild Semiconductors.

- **Perseverance**

  There are always good times and bad times. To technology start-ups, bad times appear to come very early and more frequently than good times. All the cases in this book have encountered bad times. Thus it is of utmost importance to learn to manage change when the tide is against you. Bad times can also be extremely helpful in self-review and reassessment of opportunities and the way forward. The ability to make the right choices and withstand challenges, even for just a moment longer, will determine win and loss in entrepreneurship.

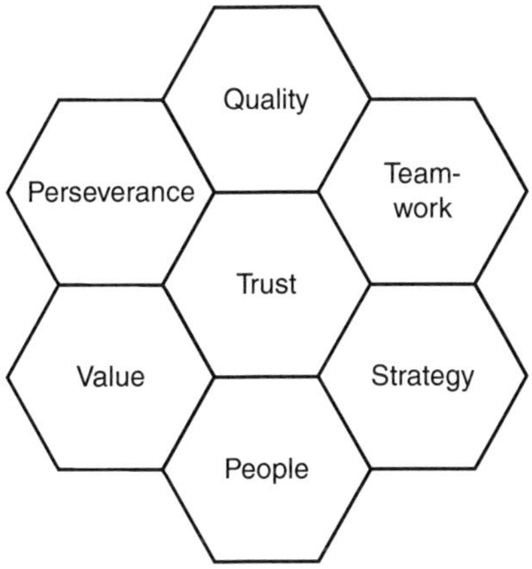

Figure 13. The Culture of Entrepreneurship: Soft Power

These values are all interrelated. A schematic diagram is shown in Figure 13.

Based on my observations, this culture holds true for new start-ups as well as established corporations in many parts of the world.

Part Four
Best Practice

Scientific and engineering research can and should be duplicable. Innovation and entrepreneurship are not. Thus research, innovation and entrepreneurship are closely related but different.

Scientific research focuses on objects in the laboratory or under predetermined conditions; innovation focuses on people and society in changeable circumstances. Entrepreneurship deals with challenges in business, markets and management in which people, the environment and technology all interact in real time and space.

In entrepreneurship, some factors are more indeterminable than determinable. The complexities encountered in entrepreneurship are much more significant than those in innovation and research, but all need the blessing of luck.

No two countries or cities are the same in the process of economic development. No system or method can be totally duplicated. Different policies, strategies and practices should be used for different cities or countries. One can learn from others, but must pursue one's own journey.

Part Four presents four cases from the last half of the 20th century: Silicon Valley, ITRI in Hsinchu, Singapore and Shenzhen. It focuses on the initial phase of the development, including the motivations, the challenges encountered and the choices made. These factors affected the future outcomes in each case. It is hoped that a country or a city can learn from these examples as references for its own development.

Chapter 16

SILICON VALLEY: THE TRAILBLAZERS

Silicon Valley is probably the best known and most studied region in the world for its ability to nurture innovation and entrepreneurship. Over a period of 60 years, it has transformed from a valley of orchards and fruits into a habitat of knowledge and high-tech business. Its success has aroused great curiosity in governments East and West. Many are interested in duplicating and emulating it as a model of economic development.

Silicon Valley is the birthplace of integrated circuits, the personal computer, internet technology, social media networks and other industries. As of 2017, it houses the headquarters of 39 Fortune 1000 companies such as Adobe, Advanced Micro Devices (AMD), Apple, Alphabet, Applied Materials, Cisco, eBay, Facebook, Google, Hewlett-Packard, Intel, LSI Logic, National Semiconductor, Netflix, Oracle and Yahoo!.[115] *According to a Wall Street Journal study, 12 of the 20 most inventive towns in America are in California, and 10 are in Silicon Valley.*[116] *The area has the highest concentration of high-tech professionals. It provides nearly a quarter of a million high paying jobs in information technology related industries.*[117]

The area also has the highest number of venture capitalists in the world to provide support to budding technology start-ups. Other professionals such as accountants, lawyers, marketing specialists, design consultants and training agencies abound. They provide the necessary support for entrepreneurship and thus constitute an integral part of the innovation system.

Although numerous books, publications and case studies have profiled Silicon Valley and its companies, the most comprehensive treatise is one co-edited by Chong Moon Lee, William F. Miller, Marguerite Gong Hancock and Henry S. Rowen titled "The Silicon Valley Edge: A Habitat for Innovation and Entrepreneurship."[118] *Discussion in this chapter makes frequent reference to this treatise in its focus on the initial phase of Silicon Valley. Knowingly or unknowingly, many pioneers laid the foundation of an innovative system in this western end of the United States. The relationships among scientific research, technological innovation and the business of entrepreneurship, discussed in Part Two of this book, are clearly demonstrated here.*

Stanford University and Stanford Research Park

Since its establishment in 1890, Stanford University has had a mission to provide service to its community. It has played an eminent role in nurturing Silicon Valley since the mid-20th century. In 1951, Professor Frederick Terman, Dean of the School of Engineering, persuaded the university to lease a part of its 8,100-acre campus to create the Stanford Industrial Park (later called Stanford Research Park) for the development of high-tech companies. This would bring technology firms closer to faculty members and students of the university. Mutual understanding among the parties would promote scientific research, research commercialization and employment opportunities for graduates. Terman also encouraged students to consider starting-up companies to pursue applications of their visions to technology. Thus, Stanford University has a tradition of turning research into innovation and nurturing students to become entrepreneurs.

In time, companies such as Hewlett-Packard and Varian Associates were formed by Stanford alumni. These start-ups along with Eastman Kodak, General Electric, Lockheed Corporation and the like moved into Stanford Park, and all have become industrial leaders in their fields.

In 1954, the Honors Cooperative Program was established to encourage full-time employees of technology firms to pursue graduate degrees at Stanford University. The participating firms reimbursed student employees by doubling what they had paid for tuition upon completion of the program. This strengthened the capability of the firms' human resources and enhanced their accessibility to new scientific advances. These measures solidified the position of Stanford University and the Research Park as the center of interaction for high-tech firms in the area.

Together, they contributed to making Silicon Valley a hotbed of innovation and entrepreneurship.

Commercialization of Research

In 1956, Terman encouraged Dr. William B. Shockley, a native of Palo Alto and an outstanding physicist at Bell Laboratories at Murray Hill, New Jersey, to return to the Valley. Shockley was preparing to establish a company in pursuit of semiconductor transistors and their industrial applications. Shockley was widely recognized for his seminal research in semiconductor physics. In 1951, he invented the junction transistor, which showed great promise in replacing the vacuum tube as the centerpiece of modern electronic devices. In 1956, Shockley shared a Nobel Prize in Physics with Bell Labs colleagues John Bardeen and Walter Bratton for their accomplishments in semiconductors.

The Shockley Semiconductor Laboratory was established at the end of 1956 in Mountain View as a part of Beckman Instruments. It planned to develop a four-layer diode with functions much superior to the current transistors. Shockley had tried to persuade some scientists in Bell Labs to come to the San Francisco Peninsula, but none would come. He thus started to recruit scientists and engineers with outstanding academic ability and technical potential from the region.

Leadership Problem: The "Traitorous Eight"

As soon as the company was established, it became apparent that the management skill of Shockley was no match for his research prowess. His leadership was autocratic and his temperament appeared erratic. His strategic focus was often confused, wavering between scientific research on solid state physics and technology developments for silicon transistors manufacturing.

One year later, eight members of the technical staff he had recruited became weary of his leadership style; they decided to leave the company and pursue entrepreneurship on their own. These members, who included Robert Noyce, Gordon Moore, Jean Hoerni, Jay Last, Eugene Kleiner, Sheldon Roberts, Julius Blank and Victor Grinich, were called the "Traitorous Eight."

This group of eight, all in their twenties, were talented, creative, independently-minded and highly motivated. Five had strong training in physical sciences. Robert Noyce, a PhD solid state physicist from MIT, was the only one with a strong background in semiconductors. Sheldon Roberts and Jay Last, both with PhDs from MIT, were a metallurgist and a physicist, respectively. Gordon Moore was a UC-Berkeley graduate, with a PhD in physical chemistry from California Institute of Technology (Caltech), and expertise in spectroscopy. Jean Hoerni was a physicist from Switzerland with PhDs from both Geneva and Oxford.

There were three engineers. Eugene Kleiner, an Austrian immigrant, was a mechanical and industrial engineer with industrial experience in the United States. Julius Blank, a mechanical engineer, had worked in Babcock & Wilcox and Western Electric. Victor Grinich had a PhD in electrical engineering from Stanford. He gained experience in transistors by working with SRI International. The "Traitorous Eight" shared a common interest: they liked the beauty, serenity and openness of Silicon Valley, and wanted to live and work there.

Turning Scientists into Entrepreneurs

Going into entrepreneurship was not the group's initial choice. As highly trained academics, they had always assumed they should be working for a company to apply their knowledge. They did not take the thought of becoming entrepreneurs seriously.

Their avant-garde proposal of planar diffusion with silicon semiconductor showed no clear evidence that it would succeed. Indeed, no major company on the San Francisco Peninsula was interested in their vision of building silicon-based integrated circuitry and making it the workhorse of electronic devices.

In great anxiety, they turned to the guidance of financial analysts Arthur Rock and Alfred Coyle. Much to their surprise, both Rock and Coyle suggested they should form a company to pursue the technology, so they started to look for investors or financial backing.

After a good deal of searching, Fairchild Camera and Instrument, headquartered in Long Island, New York, was willing to fund the project with a loan of US$1.38 million for its first 18 months of operation. It would also have the option to acquire the company for US$3 million after two years or US$5 million after eight years. Thus, a new company, Fairchild

Semiconductor, took shape. The "Traitorous Eight" became part owners individually. The formation of Fairchild Semiconductor was the subject of an excellent review by Christophe Lecuyer[119] in the book edited by Lee, Miller, Hancock and Rowen.[118]

The company was set up in Palo Alto in late 1957, dedicated to the manufacturing of high-performance silicon transistors aimed at the military sector as the first market. It quickly decided to move into another novel component: the integrated circuit (IC). Unlike discrete devices, the IC would incorporate all electronic components including transistors, diodes, capacitors and resistors, onto a single piece of silicon crystal. This would provide high reliability as well as miniaturization to electronic devices.

The company was designed to be a private sector company interested in the military sector only as users and not as research contractors. This, they believed, was needed to maintain freedom for the future of the company.

Government Impetus

During this period, the U.S. was engaged in a catch-and-match game with Soviet Russia, especially in the areas of space exploration and long-distance strategic weapons. The U.S. federal and state governments pledged large amounts of funding to support universities in higher education and scientific research. With Stanford University and the UC-Berkeley operating in this area, the Peninsula quickly rose to become an eminent region known for human resources as well as engineering and applied science. Concentrations of technical strength in electronic engineering, computer science, advanced materials, optoelectronics and computer networks emerged rapidly. Military demands for high-performance, high-quality products drove the industry to attain a higher level of excellence.

Business Growth

The decade of 1950–1960 witnessed high demand in the U.S. Air Force for high-performance digital components for avionics and control. Only silicon transistors met the reliability, miniaturization and high-temperature-performance standards of the systems. In the civil computer market, there was a pressing need for switching transistors in computer memories. This provided added factors of innovation to technology companies.

The group at Fairchild Semiconductor worked diligently to establish the planar process of silicon diffusion and other necessary system technologies in the manufacturing of integrated circuits. The list included the design of multiple components in a single chip, oxide masking and photolithography, step and repeat cameras, boron and phosphorous diffusion, gold bond wiring, silicon oxide passivation, process controls, clean room environment and many other process innovations. All of these elements became the standards of high-quality IC manufacturing and remain so to this day. The core strength of the company grew from solid-state diffusion to include expertise in optics, chemistry, metallurgy, and electrical and mechanical engineering.

At the time, the State of California was heavily populated with defense-related companies producing various products for military applications. Examples included Hughes, General Dynamics, IBM, Sperry, Rockwell, Autonetics and others. They developed digital computers, avionics control and control systems for jet aircrafts and various types of missiles.

Fairchild's IC products were very much in demand. As a supplier for Minuteman Missiles, Fairchild also had to substantially raise its capabilities in quality assurance and reliability competence. For example, the failure rate of Fairchild's first NPN transistor was 0.1% per thousand hours of operation in 1959 and reduced to 0.004% in 1961. By 1963, the failure rate dropped further to 0.00009%.[119] This represented highly challenging efforts that were richly rewarded in the market.

The Palo Alto operation was technically under the leadership of Noyes. The company showed sales revenue of about $500,000 in 1958, and quickly jumped to $21 million in 1960. The number of employees grew from 12 in 1958 to 12,000 ten years later, with an annual profit of $130 million. A manufacturing facility outside of Silicon Valley was established. The entrepreneurship has obviously met with remarkable success.

As market demands expanded, so did competition. Texas Instruments and Motorola were the closest competitors. Thus, the business growth started to show fluctuations with time. Overall, Fairchild remained the fastest growing IC manufacturer in the industry.

Emergence of Fairchildren Companies

Interestingly, all eight original co-founders of Fairchild Semiconductor left the company within the next several years. The major problem was poor communication resulting from a loss of mutual respect and trust between people in Silicon Valley and Long Island. From a larger perspective, this may be a reflection of corporate culture and of regional competitiveness between Silicon Valley and the East Coast.[122]

Fairchild Semiconductor was located in Sunnyvale, California, and its corporate headquarters was in Syosset, New York, some 3,000 miles away. The corporate executives visited the California sites once a year. Headquarters' responses to the semiconductor division were slow, few and far between. Although the semiconductor division earned most of the profit of the company, the reward was skewed and the executive staff of Fairchild Semiconductor, including Robert Noyes, were consistently compromised by headquarters.

Disheartened by this development, the co-founders and other senior staff gradually left to set up other companies or spin-offs. In many instances, they brought some company staff with them. Overall, 65 spin-off companies were formed in Silicon Valley during this period. Examples of some notable spin-offs included Amelco, Signetics, Kasper, Electroglas, Intesils and General Micro Electronics. These companies, fondly referred to as the "Fairchildren," fueled the growth of the IC industry in Silicon Valley.

Some of these companies were competing in similar fields, but as the world IC market was huge and ever-increasing, competition made the survivors stronger and smarter. Some spin-offs pursued business in IC-related areas, such as design, equipment and training. They thus formed clusters that contributed to a strong business network in the IC industry.

Some spin-offs including AMD, Rheem Semiconductor and National Semiconductor went through further merger and acquisition processes and became giants in the industry.

Formation of IC Giant — Intel

Prominent among the spin-offs was Intel Corporation, which Robert Noyce and Gordon Moore established in 1968 and soon joined by Andrew Grove. Intel developed to be the leader of IC manufacturing, outpacing Motorola and TI. It was distinguished for its x86 series microprocessor, which became the heart of most future personal computers.

Gordon Moore was famous in IC manufacturing for a "law" that commonly had his name attached. Moore's Law stipulates that the number of components (transistors, resistors, diodes or capacitors) that can be manufactured onto an IC chip will double approximately every 18 months due to advances of technology. Thus it forecasts a rapid cost reduction in the market while the performance of the chip improves. This vision had many wide-ranging implications for the semiconductor technology and business. Moore speculated that the estimate would hold true for at least 10 years. Subsequent development has shown that Moore's Law is still valid at year 2000 and has become an underlying paradigm of IC technology.

Another major impact was in the field of venture capital. Arthur Rock, who was instrumental in the formation of Fairchild Semiconductor, established the partnership Davis and Rock in 1961, which became the first venture capital on the San Francisco Peninsula. The partnership had the participation of four co-founders of Fairchild Semiconductor, who saw it as an obligation as well as an opportunity.

Eugene Kleiner, one of the original "Traitorous Eight" and an Austrian immigrant, developed expertise in the business side of start-up operations. In 1972, he became a co-founder of Kleiner Perkins Caufield and Byers (KPCB), focusing on investments in incubation and early-stage companies. KPCB was a leading venture capitalist in electronic and information technology in Silicon Valley. Donald Valentine, former Director of Marketing at Fairchild Semiconductor, soon left to form Capital Management Service at Stanford Research Park, which later became Sequoia Capital. Sequoia was to become one of the most prominent venture capitalist firms instrumental in nurturing many future technology start-ups in the Silicon Valley.

Legacy

The early phase of Fairchild Semiconductor well illustrated the culture of entrepreneurship discussed in Chapter 15 of this book. It showed how the competitive advantage of a start-up company could be fostered.

The first factor was **strategy**. The firm decided in the beginning that it would be a manufacturing organization that sells products to the military instead of a research firm seeking contracts from the military. Its profit model would be based on the core strength in volume production of high-performance integrated circuits.

It was believed this would be the **value** provided to the military and industrial sectors. Freedom from military constraints would allow the firm more flexibility in adjusting and benefiting from rapid changes in technical and market opportunities.

The eight co-founders offered different skillsets and could work in a complementary way through **teamwork**. They held the concept of **respect for people** in high esteem, and it stood as a common bond in their adventure.

By continually seeking excellence in product **quality**, the firm gained acceptance by both military and civilian customers. **Perseverance** held the group together as it went through the excruciating process of finding financial backers. If any one member quit, the adventure would have ended.

It was ultimately the lack of **respect** and **trust** between the management of Palo Alto and the management of Long Island that made the co-founders leave the company they had painstakingly established.

The entrepreneurial effort of William Shockley is also enlightening. As stated previously, scientific research, technological innovation and commercialization are related but different pursuits that require different skillsets and mindsets. It was rare to find the attributes of good researcher, good innovator and good entrepreneur in one person, which was why good team spirit among multiple talents was required. Furthermore, the spirit of entrepreneurship fell into the area of soft power, which was inadequately taught by most schools of engineering and sciences. However, soft power might have been better learned through practice than through teaching.

William Shockley never recovered from the loss of the "Traitorous Eight". He later joined Stanford University as a Professor of Engineering and Applied Science until his death in 1989. He was widely credited as the man who brought "silicon" to Silicon Valley. His greatest contribution to the Valley and to the growth of IC technology worldwide was arguably human resource development, rather than the application of research work that won him the Nobel Prize.

From IC to Knowledge Economy

The successful development of IC technology and business in Silicon Valley paved the way for the next successive innovation: the personal computer, computer networking, internet technology, social media networks, cloud computing and big data, and more recently platforms for collaborative consumption. These innovations rippled through to almost every corner of the world. This attests to the capability and power of innovation of Silicon Valley. The following list portrays the evolution of innovations starting from IC in Silicon Valley:

- IC — Fairchild, Intel, AMD, National
- Personal computer — Hewlett-Packard, Apple, Xerox PARC, SRI International, Sun Microsystem
- Computer network — Cisco, 3Com
- Internet — Netscape, Yahoo!, Google, eBay, Oracle, Amazon
- Social media — Facebook, Twitter, YouTube, Google
- Cloud computing and big data — Google, Facebook, Amazon
- Collaborative consumption platforms — Airbnb, Uber

Innovation and entrepreneurship in the green energy and biomedical fields progressed vigorously in the region. The UC-Berkeley and UC-San Francisco have played active roles in these fields. Figure 14 presents a schematic chart showing the innovation system of the Silicon Valley–San Francisco Peninsula.

Due to the rapid growth of the technology companies, the Palo Alto–San Jose area became very crowded, bringing in many of the usual urban problems such as traffic jams, air pollution and high prices. Thus, many new technology companies during the past few years have tended to move

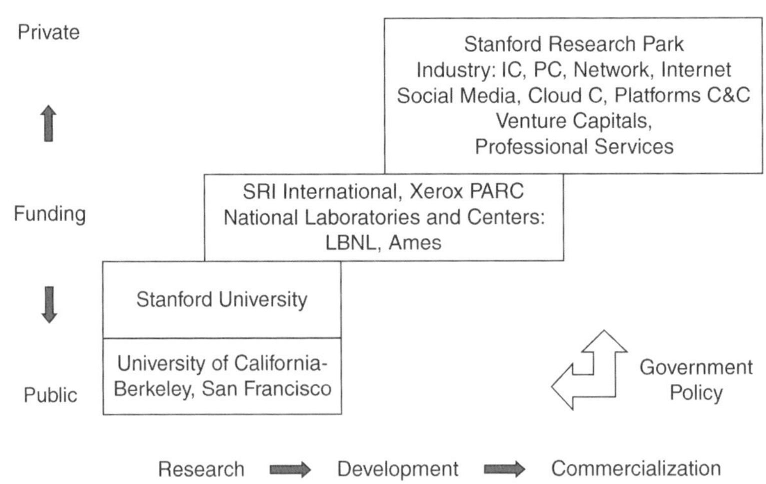

Figure 14. Innovation System: Silicon Valley–San Francisco Peninsula

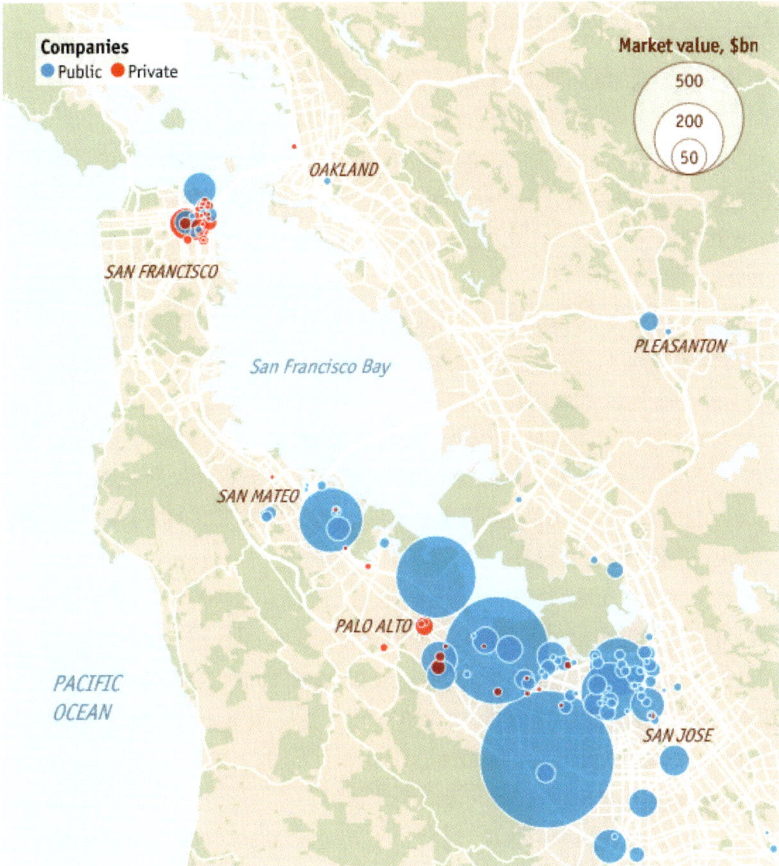

Figure 15. High-tech Firms in Silicon Valley–San Francisco Peninsula

Source: *The Economist*, July 25, 2015. Used with copyright permission.

northward, closer to the city of San Francisco and the Berkeley-Oakland area. Figure 15 illustrates the distribution of high-tech firms in the region.[120]

The city of San Francisco presently has a very high density of "unicorns," start-ups worth more than $1 billion. Among these are Uber, Airbnb and Dropbox, which have a total worth of nearly $100 billion at this point of writing. The appearance of unicorn start-ups was rather unsettling. The notion of nursing an entrepreneurship to attract ultra-large investment is quite different from building an entrepreneurship to provide sustainable value for society. Clarifying the effects and implications on this aspect of innovation and entrepreneurship appears to be a challenge for the future.

Roles of the Government

Some people have argued that government plays no role in the development of Silicon Valley. This view, I believe, is not correct. Many levels of government: federal, state and county, have been playing different roles and, at varying stages, in nurturing the various entrepreneurs, the institutions and the environment of the Valley. One would need to understand how governments work under the U.S. system of democracy.

The United States has been operating as a market-oriented economy. The government refrains from participating in any direct business activities. Through federal budget, foreign trade, foreign exchange, social security, corporate governance, and law and taxation measures, the government builds the conditions and environment necessary for economic development.

The American system is also characterized by decentralization and transparency. It affirms the rule of law and encourages open competition. The U.S. system is also favorable to new technology business ventures and does not protect hierarchies or establishments. Any business can open — and close — easily, which encourages entrepreneurship.

U.S. laws and rules apply equally with transparency in every corner of the country. How does this make Silicon Valley different from other areas such as Boston Route 128 or Austin, Texas?

In a very succinct analysis, Henry Rowen showed that federal government had helped the growth of the computer industry, and thus Silicon Valley, in three ways: as a rule maker, as a buyer of products, and as a financier and an early-stage developer.[121]

In the 1940s, the U.S. government recognized the importance of the computer to all walks of life. It encouraged innovation and entrepreneurship related to computers, but it did not support any specific company as the champion. A championship must be earned through open competition. The conditions for fair and open competition had become mature for Silicon Valley in the 1950s.

Actually, at the turn of the 20th century, the region already had excellent universities such as Stanford University and UC-Berkeley. Although Stanford is a private university and UC-Berkeley is mainly supported by the State of California, they both received substantial support from federal governments in educational subsidies and research grants.

After WWII, several federal government agencies such as the Department of Energy and the National Science Foundation furnished major and multi-year funds to research projects in many universities. The awarding of research funds was based on a rigorous vetting process, and researchers became more competitive and grew stronger as a result. In computer related fields, Stanford University and UC-Berkeley rose to become among the best in the country. More funding enabled the universities to further grow their capability in the field, train students, and nurture greater potential for innovation and entrepreneurship.

These universities hence provided a strong foundation for human resources development and as centers of excellence in research, especially in electronics engineering, computer sciences, software development and internet technology. Over the years, the government has also established major research facilities such as the Stanford Linear Accelerator Center (SLAC) and the Lawrence Berkeley National Laboratory (LBNL). These are focal points of advanced scientific research and the platforms for world intellectual exchange. Perhaps more importantly, these universities have also established close working relationships with local industries. Few academic communities in the rest of the United States can compete with Silicon Valley for outreach.

Institutional cooperation was a norm. As an example, Sun Microsystems was created by entrepreneurs from both Stanford and UC-Berkeley.

It is fair to say that no government, whether federal or state, has directly invested or participated in the development of any entrepreneurial business in Silicon Valley. However, government support has been fundamentally important in providing education, developing technology, building infrastructure, and keeping a clean and open habitat. These factors are essential for the development of any city or region. They have enabled Silicon Valley to build what some would call the frontrunner's advantage. Trailblazers like Frederick Terman, Robert Noyes and Gordon Moore fortunately appeared at the right time and place — they were able to leverage that advantage before anyone understood it.

Immigration and Citizenship

In addition, the U.S. government performed one major play related to immigration and citizenship for the benefit of everyone in the union. Silicon Valley was a frontrunner and therefore took full advantage of it.

After WWII, most of the technical and management professionals in Silicon Valley were white people who had moved in from the eastern and mid-western states. People of international origin in the region were minorities and most often took jobs with low pay levels. However, the Hart-Celler Act of 1965 had a sweeping effect that changed the landscape. The Act allowed immigration based not on small quotas according to nations of origin, but on the possession of scarce skills and on family ties to U.S. citizens or permanent residents.

For example, before 1965, a maximum of 100 immigrant visas were given to Taiwan, like most other Asian countries. Thus, in 1965, only 47 persons were allowed to enter the U.S. from Taiwan. In 1967, two years after the law was enacted, the number jumped to 1,321. The Hart-Cellar Act enabled qualified professionals from Asian countries to come to Silicon Valley to support what the new economy needed to grow. Annalee Saxenian provided an excellent review of the phenomenon.[122]

The Immigration and Nationalization Act of 1990 further addressed the needs of Silicon Valley for the future. One third of Silicon Valley scientists and engineers are immigrants. Almost two thirds of the immigrants are Asian. In fact, Saxenian reported that Chinese, Indian, Vietnamese, Filipino, Japanese and Korean immigrants accounted for, respectively, 51%, 23%, 13%, 6%, 4% and 3% of scientists and engineers. This trend continued to show in the 2000 U.S. Census.

The nature of jobs that immigrants hold in the region was also changing toward managerial and executive. From 1980 to 1998, according to a Dun and Bradstreet database, of the 11,443 high-tech firms studied, 2,001 (17%) and 774 (7%) were run by executives with Chinese and Indian surnames, respectively.[122]

Some immigrants have indeed become co-founders of technology entrepreneurships. Notable examples are:

Jerry Yang — Taiwan, Yahoo!
Vinod Khosla — India, Sun Microsystems
Andrew Grove — Hungary, Intel
Sergey Brim — Russia, Google
Eduardo Saverin — Brazil, Facebook

Highly successful immigrants took up the challenge of entrepreneurship to create jobs and opportunities for many and offered sustainable values for U.S. society and the world.

Therefore, it is perhaps no joke when someone observed that "Silicon Valley Strives on IC — I means Indian and C means Chinese."

Epilogue

On January 27, 2017, U.S. President Donald Trump, citing security concerns about terrorism, issued an executive order imposing travel and immigration bans to travelers from seven Muslim countries. This created chaos in the airports and uproars all over the U.S. In Silicon Valley, corporate leaders from Apple, Amazon, Facebook, Google, Microsoft, Netflix, Tesla, Uber and many others expressed strong opposition to it. Such a ban would hit particularly hard at Silicon Valley which relies on a flow of high level talent from India, East Asia and elsewhere. These corporate leaders see diversity as core to its identity as an innovation hub.[173]

Noting that its co-founder Steve Jobs was the son of a Syrian immigrant, Apple's CEO Tim Cook, in his letter denouncing the ban, said that the company "wouldn't exist" if such a ban had been in place. Cook concluded by quoting Dr. Martin Luther King: "We may have all come on different ships, but we are in the same boat now."[174]

It is of interest to note that multiple challenges to the order were immediately filed by individuals and organizations. On February 5, a Federal Court in Seattle barred temporarily the enforcement of major provisions of the order. On March 6, Trump issued a revised executive order for the ban but with more relaxed provisions. A Federal Court in Hawaii blocked the order before it went into effect. On June 28, 2017, the U.S. Supreme Court allowed a partial implementation of the order and scheduled a full hearing for October 2017. Quite obviously, Trump's orders have caused concern on constitutionality, among other considerations on leadership, people and trust. The utmost importance of people in innovation is a concept that political leaders can at times overlook.

Chapter 17

THE INDUSTRIAL TECHNOLOGY RESEARCH INSTITUTE IN HSINCHU: A CONSCIOUS CHOICE

When Chiang Kai-shek led his nationalist followers to sail across the Taiwan Strait in 1948, his future was uncertain. Having been soundly routed out of the mainland, he knew the challenge before him was formidable. His defeat occurred not only on the battlefield, but also in the hearts of the people.

The decade-long fight against Japanese aggression was brutal. Most of his countrymen were half-starved and fought the enemy with their bare hands. But, this did not make the people abandon him. It was when the Japanese imperialists surrendered that the real challenge began. He failed to provide the people with a decent livelihood and the opportunity to uplift themselves, and thus lost their trust.

He had very limited choices; there was no place to retreat. He made a resolution to face the challenges squarely so that he could prevail. He wanted, in his own lifetime, to implement the legacy of Dr. Sun Yat-sen, founding father of the Republic established in 1911. He would lead the government to perform its proper role, provide equitable livelihood to the people and restore their trust in him.

A New Start

Immediately after landing on Taiwan, Chiang Kai-shek worked to salvage the island by restoring roads, rails, water dams and power plants. He also undertook new measures: reforming land ownership, upgrading agricultural skills and emphasizing education. Going over to Taiwan with him were old soldiers who had been guarding the coastline, and, more importantly, were elites: engineers, managers, scholars and various business professionals. They would help him rebuild the island to show some promise for the future.

The Korean War soon broke out in 1950, and mainland China's forces were engaged in military actions thousands of miles away. With the U.S. Navy's 7th Fleet patrolling the Taiwan Strait, Chiang Kai-shek gained the much-needed breathing space and took advantage of the stability for nation building. The domestic programs were led by the elites who flocked around his son, Chiang Ching-kuo.

Education was an important long-term effort and the Constitution of the Republic of China set clear guidelines. It stipulated that for the development of education, culture and science, the central government shall spend no less than 15% of its annual budget, the provinces no less than 20% and the counties no less than 25%. As a consequence, a better-educated populace has proved to be the fundamental mover of Taiwan's social and economic progress for the next 60 years to come.

Chiang Ching-kuo had an advantage over his father. He knew what was going on outside of China in terms of technology, management, culture and politics. He had personal experience with the Russian Revolution and a keen observation of U.S.-style democracy. Most importantly, he knew that change was inevitable. As time changed, mindsets, technology and society were going to change. Thus, one needed to adjust and adapt with the change of time.

Building a Technology Economy

On the economic front, Chiang Ching-kuo knew that the meager labor intensive industry could not survive in the future. It had to undergo a major transformation and become technology intensive. He recognized that the industry was weak and the capability limited, and that the government had to play a leadership role as the agent of change. He largely delegated this role to Y.S. Sun, an electrical engineer, K.T. Li, a physicist, and S.S. Shu, an applied mathematician.

All three had worked at the cabinet level at different times and in different capacities, and had made immense contributions to the rise of Taiwan as one of Asia's Four Little Dragons. Among other accomplishments, they laid the foundation of the ITRI and the Hsinchu Science-based Industrial Park (HSIP), which would become the cradle of Taiwan's technology economy.

Sun, as Minister of Economic Affairs, pushed forward legislation to establish the ITRI as a national institute in 1973. Later, in the late 1970s to mid-1980s, he was appointed Prime Minister and led the Big Ten National Construction Programs. These programs successfully upgraded Taiwan's infrastructure including: the North-South Trans-Island Highway, railroads, seaports, the international airport, nuclear power plants, advanced steel making, ship building, petroleum refining and petrochemical development. Completion of the Big Ten National Construction Programs bolstered Taiwan's economic foundation and enabled it to go forward.[123,124] As Sun moved on to high technology, the ITRI was his major vehicle.

Li had served as Minister of Economic Affairs and Minister of Finance. Respected for his foresight and seniority, he established a good professional network and was known for his capability of implementation. In the Sun cabinet, he was appointed State Councilor on Science and Technology, with the function to coordinate different government agencies in matters related to science and technology. He had previously facilitated the development of export processing zones and industrial parks, and later focused on applications of information technology to government functions, legislation to assist initial public offerings, venture capital and major national science and technology initiatives.[124]

Shu, president of the National Tsing-hua University (NTHU) and later Chairman of the National Science Council, initiated programs to support university research and large-scale R&D projects, the return of science and engineering experts in the U.S. and Europe and scientific exchanges between Taiwan and the West. A man of exceptional vision, he advocated the development of geothermal energy, solar energy and electric vehicles, all of which were perceived as avant-garde in the 1970s.[125,126]

In 1975, Shu proposed establishing a science park in Hsinchu for foreign and domestic companies to establish R&D centers and to nurture technology business ventures. This required major government investments in infrastructure: land, electricity, water, gas, communication and construction. Heated debates on the feasibility of this project occurred in ministerial meetings. Ultimately, opposing opinions centered on affordability. Although the benefits were uncertain, the costs were real and immediate. At a time when government resources were limited and the burden heavy, the debate was quite understandable. However, sharing his vision for the future, Chiang Ching-kuo fully supported Shu's proposal.

Chiang and Shu did have different preferences on the location of the park. Chiang preferred a site at Long Tan near a large research institute for defense technology. Shu argued that the park was going to focus on civilian industry and facilitate private sector participation. Thus, he believed that a site away from a military setting would be better. Furthermore, Hsinchu was going to have the technological support of the ITRI and the two national universities. This was of course before "national innovation system" became a popular term. Finally, Chiang agreed.

The HSIP project was formally approved in August 1976. Construction did not start until late 1978 when legal rights to all of the lands were secured. At the grand opening two years later, on December 15, 1980, few cabinet members participated in the celebrations. Chiang led the ceremony and made a landmark speech on the future of Taiwan's industries and the role the HSIP would play in fostering that vision.[126] After the ceremony, Chiang and Shu were seen touring the park riding on an experimental electric vehicle (EV), the Tsing-hua EV-5, the prototypes for which Shu had been instrumental in developing. When I served at NTHU as Dean of the School of Engineering in 1978, I led the EV program, succeeding my college classmate Kao-wen Mao, who had started the program at the university two years earlier.

Forming an Innovation System

Unknown at the time, the park would become a technology business incubator, not so much for foreign companies, but for domestic entrepreneurs. The government originally hoped to have big-name international companies come to the park as anchors. However, few were really interested in the prospect of Taiwan. Luckily, as Shu had designed it, the park neighbored the ITRI and two national universities, NTHU and NJTU. It developed to be an excellent proving ground for innovation and entrepreneurship arising from these organizations. While new technology ideas were being tested, researchers, faculty members and students shuttled in and out. Thus, 20 years later, a new landmark in technology and innovation was

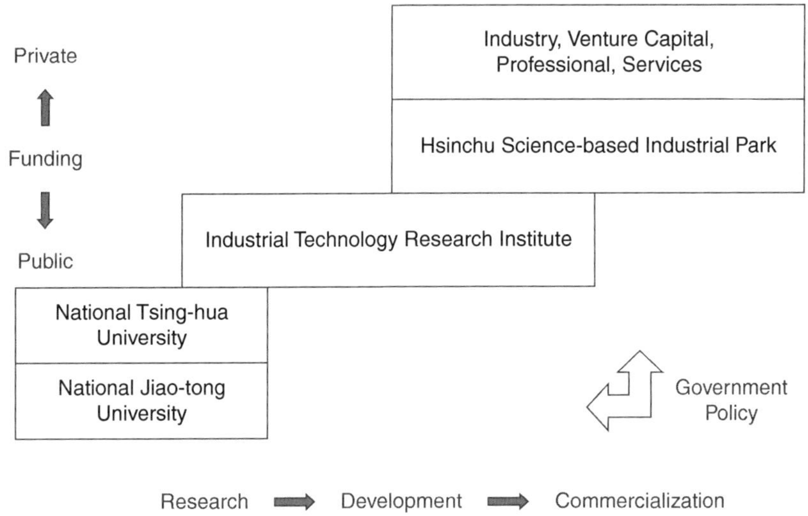

Figure 16. The Innovation System of Hsinchu

Source: O. C. C. Lin, 2002.

established for the small city of Hsinchu, previously known as the Windy City and famous for fine rice noodles and exquisite meatballs.

The innovation system, or eco-system, of Hsinchu is illustrated in Figure 16. The x-axis represents the innovation process, from research to development and commercialization, and the y-axis represents the sources of funding, from public to private. Playing a central role in Hsinchu's eco-system was the ITRI, which was chartered to advance Taiwan's industrial technology.

When Sun proposed the legislation for the establishment of ITRI, there was strong sentiment against it in the legislature, as some government assets were to be endowed to the new institute, a private entity. After a lot of consultation and debates, Sun finally won the necessary support and the ITRI was established in July 1973 as a non-government organization. It received some government funds for its establishment, but was to operate as a corporation and responsible for its own existence. The institute would be led by a president as the chief executive, who would be supervised by a Board of Directors. The board members would include government ministers, university presidents and business leaders. The prime minister would appoint both the chairman of the board and the president of the ITRI. This was the duality in the ITRI's nature: a national laboratory and an NGO.

At its formation, the ITRI consisted of three government laboratories that were then privatized to join the ITRI. The government would present the land, personnel and assets, such as the building, equipment and budget the laboratories possessed, to the ITRI. The personnel, who were formally government officials, became private citizens. The laboratories separately focused on mining, metallurgy and chemicals. Their major functions were originally to provide technical support to government agencies in implementing governmental regulations and practices. There were little or no R&D activities. Thus, at the beginning, the ITRI could be quite accurately characterized as an organization standing on very thin ice.

As it strived to adjust to its new role in a rapidly changing environment, ITRI, in its first 10 years, was somewhat lackluster. The only significant development was the preparation to move into a new strategic area of microelectronics.

With the Big Ten National Construction Programs nearing completion, Taiwan's economy gained new strength. This provided the foundation for Sun's government to embark on new national science and technology initiatives. Despite a major diplomatic setback due to the severing of its formal diplomatic relationship with the U.S. at the end of 1979, Taiwan headed into the 1980s with a new sense of urgency and self-reliance. The ITRI was positioned to play a major role in the national development of thrust technologies at this important juncture.

The 20 years after 1980 were exciting as the period of rapid growth for the ITRI and for the development of Taiwan.[81a,81b,149] In the following sections of this paper, I describe the major technological and managerial developments which have shaped the ITRI as an institution of innovation and its impacts on Taiwan's economic growth.

Focus on Electronics

Witnessing the rapid development of information technology in the U.S., Sun felt the need to set up R&D in microelectronics to nurture a domestic electronic industry. Following the advice of W.Y. Pan, Sun authorized a technology transfer proposal involving the purchase of a wafer fabrication line from the Radio Corporation of America (RCA) to be established at the ITRI as a centerpiece of industrial research. At the time, the president of the ITRI was H.C. Fang, who was formerly Vice Minister of the Ministry of Communication. Fang quickly organized a team under the leadership of D. H. Hu, a professor of electronic engineering at the National Jiao-tong University, to undertake this project.

The RCA fabrication line could produce 7-micron complementary metal-oxide-semiconductor (CMOS) integrated circuits on 3-inch silicon wafers. In 1974, this technology was really outdated even for the RCA. Thus, the transaction was of no concern to the U.S. government. The political environment in Taiwan was focused and forward-looking, and the leadership was dedicated and communicative. Thus, the relative obsolescence of the fabrication line was of no objection to any vetting committee. All agreed that this was a good choice as a starting point in terms of cost and benefit.

A team of approximately 30 engineers was sent to the RCA's factory in 1976 to receive training, learn the operations, dissemble the equipment, carefully pack it and ship it to the ITRI. Upon arrival at its new home in Hsinchu, it was reassembled and received the most adoration, attention and care from the ITRI engineers who had previously only seen schematic diagrams of IC fabrication in textbooks.[127]

The Electronic Research and Service Organization (ERSO) laboratory was formally established in 1979 as a division of the ITRI. Hu was appointed Director and supervised various projects around it. In a few years' time, numerous improvements were made, which gave the ERSO the ability to fabricate 3.5-micron CMOS ICs.

In 1980, a group of professionals left the ITRI with a technology transfer license to pursue the commercialization of CMOS technology. The company, the United Microelectronic Corporation (UMC), was led by Robert H. C. Tsao, Deputy Director of the ERSO. The UMC was the first IC company in Taiwan. The group became the first entrepreneurs to explore a major technology developed by the ITRI.

In the subsequent years, many start-ups formed to pursue business related to IC: design, testing, packaging and systems integration, all fabless by itself but complementary in the IC business value chain. Most RCA team members and early ERSO project participants left the ITRI to take leadership positions in these start-ups. Quite obviously, the experience and expertise developed at the ITRI for young professionals became highly attractive and marketable to the industry. Thus, they were natural targets for corporate technology leaders.[127] They were the new breed of business leaders in Taiwan, and many became millionaires. These alumni afforded the ITRI a certain unexpected fame.

The movement of experienced professionals from the ITRI to the industry catalyzed a new wave of private investments in new technology fields in Taiwan. These activities further aroused the interests of many experienced professionals in the U.S. and Europe. Many returned to Taiwan to participate in private business or to undertake entrepreneurial endeavors. This resulted in the formation of clusters of technology companies in semiconductors, electronic materials, computer software, communication devices and information technology applications. All contributed to the booming information technology industry in Taiwan. Naturally, the majority of companies were located in the HSIP.

In mid-1985, Morris Chang, President of General Instruments, was recruited as the president of the ITRI after Fang's retirement. Chang was well known in the U.S. electronics industry, having served in senior executive capacities at TI. His coming to Taiwan strengthened the position of ITRI in the global electronics community.

National Science and Technology Initiatives

In the mid-1980s, Premier Sun invited a group of distinguished scientists and business executives from the U.S. and Europe to form the Science and Technology Advisory Group (STAG). The advisors would provide guidance, based on their understanding of the global landscape of science and technology, for the future development of Taiwan.[128] The premier also convened the National Science and Technology Conference to help map strategic objectives. STAG advisors participated in the conference, along with government, university, industry and technology institute leaders. These were openly publicized events in Taiwan and were closely watched by many economic entities in the Asia Pacific region.

Over a six-year period, the following fields were identified as major objectives, referred to as the national thrust objectives: energy, materials, information, automation, optoelectronics, precision machinery, food processing, hepatitis prevention, industrial pollution control, ocean science and technology and synchrotron radiation.

Development proposals were requested from all institutes and universities with varying degrees of industrial participation. Through expert committee vetting, qualified proposals would be included for basic funding in the annual budget of relevant government agencies. After approval of the budget by the legislative branch of the government, the institutions undertook the projects for implemention.

This became a basic model of strategic planning, budget appropriation and project execution in the following decades for national projects. In the later years, the process was tainted by a great deal of haggling and politicizing by the legislators involved in the legislation process.

The ITRI Enters Its Growth Phase

Six national thrust objectives (information, materials, energy, optoelectronics, automation and pollution control) fell within the ITRI's basic strength. They represented a great opportunity for the ITRI to further develop its capabilities and to contribute to national needs. The funding was largely under the purview of the Ministry of Economic Affairs (MOEA). Thus, the ITRI fostered a strong partnership with the MOEA and the industries it served. The results of these national thrust projects further bolstered the development of the innovation system in Hsinchu and in Taiwan.[81a,81b] Some key developments in each field in the initial phase are highlighted here. More details and updates are provided on the ITRI, National Science Council (NSC) and MOEA websites.

Microelectronics was the centerpiece of information technology, which was one of the first groups of national thrust objectives. With MOEA support, the ERSO continued to pursue R&D in microelectronics under the directorship of Chintay Shih. In time, the technology scope considerably expanded to include computers, communications, computer peripherals, software development and network devices. By 1987, the microelectronic projects had accomplished the fabrication of 2-micron CMOS ICs on 6-inch silicon wafers. Another spin-off was being considered at that juncture.

Materials science and technology was the area in which I had a specific responsibility. I joined the ITRI in 1983 as General Director of the Materials Research Laboratories (MRL). The MRL was established to assist Taiwan on two levels: upgrading the competitiveness of its industry by making use of modern materials technology, and, establishing new industries in materials-related fields.

I felt strongly for the plight of Taiwan's small and medium-sized enterprises (SMEs). They provided most of the employment and represented the mainstream of the Taiwan economy but were in a dire need to upgrade their competitiveness in the world marketplace. Although they were enthused about the arrival of high technology, their survival hinged on raising efficiency and productivity of current products. Therefore, two-pronged objectives were needed.

After reviewing global trends, the status and potential future development of Taiwan's industries, we decided to expand the scope from traditional metallurgy to include high-strength alloy steel, compound semiconductors, polymer composites, engineering ceramics, corrosion controls, materials analysis, characterization and selection. With MOEA funding, we strived to make the MRL a center of excellence in materials science and technology and an active partner in upgrading Taiwan's industries.

Within four years, the MRL completed many technology transfers. An example was the spinoff of electro-slag remelting technology to form a company producing high-strength alloy steels (program leader: Robert H.S. Chen). Another example was providing fiber composite technology to the sporting goods industry and, more specifically, to a company to establish a new brand of high-end carbon fiber epoxy composite bicycles (program leader: Roger J.X. Liu). This company later became number one in its field in the world market (Chapter 20). The corrosion project helped the Taiwan Electric Power Company improve operational safety and reduce costs (program leader: T.J. Chung). A project to develop flexible printed circuit board materials was undertaken to support the development of future electronic devices with new values: smaller, thinner and lighter. The program leader was Z.M. Liu, who was later promoted to General Director of MRL and subsequently, president of ITRI.

Commercialization in microelectronics research occurred at the end of 1987. The ERSO's very-large-scale integration (VLSI) ICs on the fabrication of 2-micron CMOS ICs on 6-inch wafers was targeted for commercialization. Due to its complexity and the large capital investment, approximately US$300 million, Morris Chang decided to lead the commercialization program. A great deal of effort was dedicated to organizing the investment group, which consisted of 10 local industrial companies, Philips Electronics and a government-supported development fund. The Taiwan Semiconductor Manufacturing Corp (TSMC) was thus established with Chang as president. It was the first IC foundry in Taiwan and became the largest in its field worldwide. Today it occupies approximately 45% of the world IC market share.

I succeeded Chang as the president of the ITRI in 1988. Having spent five years with the ITRI, I have concluded that, for ITRI to maximize its impact in the future, it needs to become an organization with innovation capacity, entrepreneurial mindset, corporate culture and modern business processes. In other words, the institute needs to undergo organizational improvement to support technological development. Whereas technological development is the responsibility of each laboratory, organizational improvements rest in the headquarters which are the president's responsibility.

In practice, I placed focus on four areas. The first was to ensure that the institute operate with proper business processes in human resources, financial management, strategic planning, review and appraisal, and technology diffusion. The second was building a corporate culture in innovation and entrepreneurship so that members of different laboratories and backgrounds could take ownership of the institute's mission and commit to its objectives. The third was broadening the ITRI's technological capacity, not only to nurture new and high technology industry, but also to bolster the existing SMEs' industries by upgrading their capability in technology and innovation. This, in essence, was under the vision on how the future Taiwan industry can be revitalized on technology along the two-prong strategy of development.[81b] Finally, the fourth was improving the physical infrastructure by building a 120,000 square-meter "Innovation Plaza" on the Chu-Tung campus.

Financial policy was an important issue. Chang and I both felt that the duality of the ITRI was an important asset for its sustainability. Therefore, the ITRI would seek to obtain government funding only for long-term R&D projects targeting at emerging industrial technologies. The institute would work with private sector businesses and public-sector agencies for technology upgrades and problem solving, thereby supporting their businesses and operations. Such activities would be funded by clients. We established a "One-to-One" guideline of government research funding to industrial services income. This one-to-one policy became a guiding principle for project planning and execution at all ITRI laboratories. Conducting major communication and training programs for all managerial staff for their commitment to the new ITRI culture and business processes became a priority of the top executive team consisting of the President, two Executive Vice Presidents: Chintay Shih and Anthony Ku, and all the General Directors of the Laboratories.

Industrial automation was a major national thrust program with the objective to promote the automation of Taiwan's industries to achieve higher efficiency and productivity. Taiwan's industrial structure was very complex and consisted mainly of SMEs. They did not have the knowledge, tools or training to embark on automation. Thus, assisting tens of thousands of small companies was a very difficult, if not impossible, challenge.

We recruited Richard C. Liu, a professor of industrial engineering at Purdue University, to be the General Director of the Mechanical Industry Research Laboratories (MIRL) in 1990. After studying industry needs, a strategy was laid out. To achieve industrial automation, we would first establish an automation industry and leverage it as the agent of knowledge transfer.

We sought industrial partners to develop the various components of automation technology, such as designs, sensors, controls, actuators, motors, programmable control devices and production system integration. After our partners became well versed in the technology, they moved forward to commercial operations. They would be the agents to promote automation by showcasing their products to any target industry clients. This proved to be the right choice for Taiwan. In the later years, the machinery industries increased greatly its added value, providing support to the growth of all the Taiwan industries downstream.

The MIRL also conducted several significant programs, such as those focused on clean motorcycles, automatic warehousing and industrial robots, starting when C. M. Shyy was the director.

Supported by an MOEA grant, the MIRL led an industrial consortium in 1991 to collaborate with Lotus Engineering in England on the design of a 2,000cc automotive engine. With the China Motors Group and Yue-Loong Motors as partners, the prototype engine was quickly produced for commercial use. The MIRL was the first in Taiwan to design and manufacture the powertrain for a selected class of automobiles.

Energy was among the first group of national thrust initiatives. It was a pressing issue in Taiwan, almost all of its petroleum and fuels were imported.

In the early 1980s, the Energy Laboratory was built on the foundation of mining and minerals research as core activities. Under the directorship of Hsuan Yeh, a professor from the University of Pennsylvania, it established an objective to develop solar energy. Being a novel concept at the time, the Taiwan industry was not ready for it. Thus, only a few experimental programs in photovoltaic technology had been carried out, which later became the seeds of Taiwan's solar photovoltaic industry.

A few years later, Yeh left the ITRI and the general directorship was succeeded by Robert J. C. Yang, a professional energy consultant from the U.S. After I assumed ITRI presidency, the laboratory was reorganized as the Energy and Resources Laboratory (ERL). Experimental programs were established in energy conservation, treatment and conversion of biomass, incineration, green energy, and energy auditing, consultation and services. It offered broad based support to all the industries in Taiwan. In the later days, it was also active in carbon-neutral initiatives.

Optoelectronics was a growing field and the ITRI had basic resources at the Union Chemical Laboratories (UCL). It was capable of growing mono-crystalline compound semiconductors (Chapter 21). With this foundation, the MRL enhanced its capability from synthesizing single crystals to epitaxial films by establishing the molecular organic chemical vapor deposition (MOCVD) technology. This opened up the possibility of developing a range of light-emitting diode and laser diode applications.

Recognizing its potential, I expanded the research group to become Optoelectronic Research Center with Ming-Hseung Lin, Deputy Director of MRL, as General Director. After I assumed the ITRI presidency, the center was upgraded to be the Opto-Electronics and Systems Laboratory (OES) and became a full division of the ITRI.

The OES further developed its capabilities in the electronics and optics fields. It worked with the industry to develop non-spherical lenses and devices, hard disk drives and computer peripherals, such as integrated photocopiers, faxes and scanners. It also moved into the precision medical devices area. Eric Lean, a senior member of IBM Research, later succeeded M.S. Lin as General Director and further led the laboratory into many advanced technology areas.

Industrial pollution became serious for Taiwan as industrialization proceeded. Because of the complex industrial composition, Taiwan faced a collection of complex and widely-spread pollution problems. For example, electrical and electronic factories produced chemical wastes with metallic contamination; domestic animal farms produced gaseous waste and biomass; food processing works produced solid waste and liquids; and textile dye-stuffs plants produced liquid waste and toxic organics. Although chemicals were the common denominator, abatement technologies were different and existed in different laboratories. They needed to be coordinated. Thus, the Center of Environmental Protection was established. L. Y. Chen, Deputy Director of the UCL, was its first General Director.

Industrial safety and health was a growing concern in Taiwan. Unlike pollution problems which were usually continous and gradual, safety problems could be abrupt and catastrophic. Furthermore, fires, explosions and chemical hazards generally occurred together with disastrous consequences. When the public and government recognized the disasters, it was usually too late for remedy. Recognizing the importance of the issue as Taiwan became more industrialized, the ITRI established the Center of Industrial Safety and Occupational Health (CISOH) in 1990. Ada Wen-shung Ma, experienced biochemist and Chief Industrial Hygienist at the Philadelphia Naval Base, was its founding General Director.

As it developed, the center was known for expertise in fire, chemical and explosion hazards prevention. It also specialized in HazOp and related safety auditing technology. In 2007 it was spun off from the ITRI to become a separate entity, the Safety and Health Technology Center (SAHTECH) to provide consultancy services in industry safety. Most of the companies in the HSIP and other science parks became clients of SAHTECH (Director: R.B. Yu). Furthermore, many former members of CISOH were recruited as chief industrial safety officers by Taiwan companies whose operations spread over the entire Asia Pacific region.

Computer and communication were not in the original list of national thrust objectives. It emerged from the information-microelectronic program. With VLSI projects underway, the use of ICs naturally spawned into the computer and communication fields. Soon the rapidly growing needs for design, testing and service in these areas became a distraction to our highly focused IC program. We realized that the demand and development of the computer and communication industries in Taiwan would soon overwhelm those of the IC industry.

After much deliberation, we decided in 1991 to separate the ERSO into two laboratories. One was to continue focus on ICs and carry the ERSO name. The other, the Computer and Communication Laboratories (CCL), would be dedicated to these new fields. We also recruited Steve R. Y. Cheng from the Bellcore Corporation to be the General Director.

The CCL rapidly became a star among the budding industries in Taiwan. In 1990, Cheng proposed, shortly after he was on board, to develop a notebook-sized, portable personal computer (PC). I was unsure of the market needs of this novelty and wanted industrial partnership in the program. In conjunction with the Taiwan Electrical and Electronic Manufacturers Association (TEEMA), Cheng called a meeting with the industry. He announced the proposal and solicited partnership to form a technology development consortium. Eash partner was expected to pay a membership fee of approximately NT$1.2 million (approximately US$40,000). This sum of money was not large, but neither was it a sum that industrialists would part with lightly. There were over 30 applications on the spot. We finally settled on a membership fee of NT$1.8 million. It did not discourage participation one bit. Ultimately, the consortium had 46 members. It was the largest institute-industry alliance for new product development.

The first laptop PC was completed within the one-year target. It made its debut at the 1992 Consumer Electronics Show in Las Vegas and attracted a great deal of attention. As promised, a schematic diagram with basic criteria of the laptop PC was given to each member. The ITRI also provided training on the basics, but it was understood that any design and manufacturing of commercial products was solely the member's responsibility. Approximately one third of the participants eventually went into commercialization.

Taiwan's market share of laptop PCs in the world grew steadily from that point. In 1992, its world market share was less than 1%. It grew to approximately 32% in 1996, 50% in 2001, 86% in 2007, and nearly 90% in 2011. Most Taiwanese PC manufacturers pursued collaboration with brandname manufacturers worldwide to take advantage of the established market channels. As IDMs, they could fully make use of their technical advantage. Most PC companies that had their base in the HSIP later spread to the Chinese mainland as the relationship between the two sides of the Taiwan Strait relaxed.

Another important effect of the CCL/ITRI was the diffusion of human resources. Similar to the IC industry, many CCL team members involved in laptop PC projects left in subsequent years to join other IT companies as vice-presidents, presidents and chairmen.

The notebook PC consortium was considered an efficient and effective model of innovation. This model was used many times in other ITRI projects. It was really a reflection of how a good innovative eco-system could promote the growth of a technology-based economy.[62]

A Snapshot

The HSIP became home to approximately 550 technology companies over an area of 650 hectares in Hsinchu. The four largest business groups were IC, optoelectronics, computer peripherals and telecommunications. They constituted approximately 75% of all companies. The rest were precision machinery, biotechnology and other businesses. At present, the park has the highest density of 12-inch wafer fabrication lines in the world. The average R&D expenditure of companies in the park was approximately 7% of their sales revenues. The government has since established industrial parks in Tainan and Tai Chung modelled after HSIP.

The ITRI continued to strive in its field of focus. Since 1994, the presidency had been succeeded by Chintay Shih (史欽泰), Johnsee Lee (李鍾熙), Jyo-ming Shyu (徐爵民) and, presently, Zhong Ming Liu (劉仲明), all rising from the staff with unswerving dedication and proven records. The various laboratories were reorganized several times to adapt to the change of the business environment. The core capabilities remained as: electronic and optoelectronic systems, information and communications, mechanics and mechatronics, materials and chemicals, and green energy and the environment. Biomedical technology and devices were added after 2000. It grew from the strength of the UCL, OES, CCL and MRL.

The ITRI's professional personnel had stayed at approximately 6,000 over the recent 20 years. It established close connections with local industries and many global technology institutions. The ITRI had the capability, research infrastructure and corporate culture to move rapidly into any new technology field.

The ITRI had established approximately 240 technology companies in Taiwan over the years. Although most of these companies grew from brand new start-ups, a good portion of them were branched off from existing companies to pursue new technologies. This diversity gave Taiwan's industries new vitality to face the challenges of the new millennium.

An estimated 18,000 ITRI employees have left or retired early from the institute over three decades. Most have joined the industries and assumed leadership roles in Taiwan industry. They are change agents for upgrading Taiwan industry, in both technological and management capability. The ITRI culture has established role models for the next generation Taiwan industries (see "Ardentec Corporation", Chapter 22 of this book).

By 1995, Taiwan had become the third leading information products producer in the world, next to the U.S. and Japan. The total output value of electronics, machinery, chemicals and transportation vehicles in 2003 reached NT$4.77 trillion (approximately US$160 billion).[151]

The ITRI has received wide global recognition. In 2014, Thomson Reuters rated the ITRI one of the top 100 most innovative organizations in the world. In the year 2016, it received five R&D 100 Awards.

Outcome: Sustainability of Taiwan

1. Taiwan's successful transition into a technology-intensive economy started in the 1990s. The transition was attributed to timely government actions under Chiang Ching-kuo's leadership. One of these actions was the establishment of an eco-system to promote technology and innovations in which the ITRI and HSIP played central roles. The system raised the industry's technological capacity and nurtured new entrepreneurship. In the last quarter of the 20th century, Taiwan's achievement was remarkable and had a high global competitiveness ranking.[63,148]

2. The ITRI maintained its defined position in the innovation system. It focused on the development of industrial technology and connected the academia and the industry. It supplemented the work on both ends, but did not duplicate or replace either. During the early 1980s when Taiwan's academic research was weak, ITRI worked extensively with international corporations and institutions. It functioned as an agent of globalization for Taiwan's economy. The criteria of the success of ITRI programs included: number of technologies transferred, industrial value created, nurturing industrial leaderships, cumulation of patents and IPR and technology consultation for public agencies and private sector industries.

3. The ITRI's contribution to Taiwan's economy through its spinoff companies and technology recipients was estimated to be 10–12% of

the annual GDP.[62] The ITRI was recognized as one of the world's top 100 innovative organizations.

4. SMEs and new start-ups were the direct beneficiaries of the ITRI technology transfer. Thus, technology benefits were distributed to the middle and lower echelons of Taiwan society. This, in effect, contributed to social harmony by reducing the economic divide and polarization of the society. The growing middle class contributed to Taiwan's political stability. Consequently, the government was able to implement political as well as economical transformation. Through universal suffrage, the democracy process also proceeded well in Taiwan. After Chiang Ching-kuo, there has been four successions of the presidency through peaceful general election.

5. On an individual level, the ITRI's approximately 18,000 alumni led Taiwan's industries at various executive and management capacities. They were leaders in innovation and entrepreneurship in Taiwan. In addition, several later served as cabinet-level ministers and university presidents.

6. The ITRI played a pioneering role in promoting technology and innovation exchanges with the Chinese mainland. In January 1994, Prof. Guan-chao Zhou, President of the Chinese Academy of Sciences, led a group of 25 leading scientists and engineers to visit the ITRI and participated in scientfic and technology conferences.[81c] This extraordinary visit was in response to the 1993 visit of the ITRI President to CAS. Exchanges at this high level were considered as ice-breakers to help subsequent technology and business exchanges between the two sides, despite the complex political environment across the Taiwan Strait.

7. After the year 2000, fierce internal political rivalry took place in Taiwan. The rising struggle with the Taiwan Separatist Movement had caused polarization in social harmony, deterioration of cross-strait relations and a sluggish economic growth. It should be alarming to note that the ratio of GDP between China and Taiwan in 1993 was 2 to 1 and, in 2013, it had jumped to 20 to 1. Taiwan's position in competitiveness of economic development was significantly eroded in the flight of time. The roots of the economic tree were shaken due to shifting political vision and indecisive governmental priority. Critical choices are needed at this historical juncture for Taiwan to stand up to the monumental challenge of sustainability.

Chapter 18

SHENZHEN: FORMATION OF A CITY OF ENTREPRENEURSHIP

In 2016, Shenzhen, adjacent to the northern border of Hong Kong, was the city with the highest GDP per capita in China. Over 50% of the economy was high-tech related. This was an exceedingly remarkable accomplishment considering 30 years ago, the city was merely a fishing village. The city now drew its economic strength from technology and entrepreneurship. Indeed, budding start-up companies were spawned in the city's many technology parks and virtual campuses.

What made this accomplishment possible? To understand it, one needed to review the big environmental changes that had shaped the city in the previous three decades.

The Time

In 1957, Mao Zedong sought to enhance China's industrial capability so that within 15 years, China could overtake Great Britain in steel production. And, in another 15 years, China would be able to compete directly with the U.S. Thus, Mao catapulted the "The Great Leap Forward" movement, which created a frenzy for homemade steel mills in the backyards of every village around the country. It was suggested that lacking the technology and supporting infrastructure was not important as long as Mao's political thought was upheld. This had created disastrous effects on the country for generations to come. After Mao's death, the desire for economic growth continued to be a conscious motivation for China. However, new approaches were adopted.

A major change for China occurred in the early 1980s under the leadership of Deng Xiao-ping, who set out to lead China in pursuing social and economic reform. Gone was the sloganeering of the Cultural Revolution (1966–1976). Science and technology were affirmed as the first factor of productivity, and scientists and engineers were no longer treated as enemies of the state. Also, universities were reopened to nurture human resources which are essential for the national modernization efforts.

Determined to build a socialist society with "Chinese characteristics," Deng took a pragmatic approach, leaving economic and political doctrines to the side. There were two key strategies. One was often referred to as "black cat or white cat, whichever one catches mice is a good cat" dogma. In essence, the government could and would take any action deemed appropriate to build the country and not be bounded by political ideology. The second could be viewed as "crossing the river by feeling the stones". This was experimentation with caution. In essence, China decided to learn and try anything stepwise to establish an industrial economy and quickly build its national strength through modernization.

As for implementation, the central government allowed certain decision-making authority on industrial projects to trickle downward. This encouraged and emboldened the provinces to take initiatives under the framework of the national Five Year Plan. With the size and population of most provinces in China ranking comparable with most OECD countries, the scale of effort was awesome, the compounding effect was massive and the total effects on the national economy were unmistakable.

In June 1989, an unfortunate event occurred in Tiananmen Square in Beijing that resulted in the deaths of many young students participating in reform movements. The incident created a shockwave around the world and left China in isolation for a number of years. However, by firmly upholding the reform and modernization policy, Deng provided China the much-needed stability. By 1995, the world outside had busily exploited the arrival of information technology, especially the internet. In China, institutions of higher learning had been reopened and the people showed growing interest in the global community.

The period between 1995 and 1999 was an exciting one in the history of technology and innovation. Many young people in China became increasingly aware of world events and would fancy positioning themselves in the globalization era.

Looking to the U.S. as the trailblazer of innovation, they were fascinated by the creation of Yahoo.com in January 1995, which showcased a highly versatile search engine and a popular domain in the cyberspace. Its co-founder and "Chief Yahoo," Jerry Yang, was a young graduate of Stanford University. Born in Taiwan of Chinese origin, he went to the U.S. as an immigrant at 10 years of age. Featured on the covers of *Time* and *Business Week* in 1998, Yang became a star to millions of youth his age in the Greater China Region.

Then came the spectacular success of Netscape, a young start-up company that specialized in web browsers. On August 9, 1995, its stock, which had been set at US$14 per share, soared to US$75 at closing, giving Netscape a market value of approximately US$3 billion. This signaled the beginning of a new industry and a new era. The strong interest of information technology giant Microsoft in internet and technology and the subsequent corporate maneuvers regarding Netscape dominated the news. In 1998, America Online acquired Netscape in a US$10 billion merger.

In 1998, Steve Jobs debuted his iMac after returning to Apple Inc., the company he had founded but was later forced out. The new device created an exciting horizon for people worldwide.

In the mainland details of these events were largely unknown to the public. But the landmark news captured the fancy of many ambitious young and educated minds.

In 1995, Ma Yun in Hangzhou set up a company called China Yellow Pages to offer directory services. Ma hoped to help SMEs engage in import-export trades through the internet. In 1999, he formed Alibaba.com as a business-to-business platform to facilitate electronic commerce for the SMEs.

In 1998, Sohoo.com was founded by Chao-yang Zhang, a graduate of MIT. The site took after the Yahoo model and was later changed to Sohu.com. This was followed by Sina.net, established by Zhi-tung Wang, and NetEase, established by Ding Lei, who operated the e-mail domain 163.com. These were three leading internet companies that offered software, news, e-mail and other online services. In a relatively short period, they gained national fame. All of these companies' leaders were from the post-1960s and post-1970s generations.

In the search engine area, Baidu was formed in 1999 by Robin Li, who graduated from Stanford University and returned to China to pursue the venture. Baidu quickly became the first search engine in the Chinese language. In other areas such as computer games and travel services, new companies sprung up everywhere, as market demands have been gradually stimulated.

The increasing interest in innovation and entrepreneurship in China paralleled the advent of internet technology. This was unprecedented, as the new wave was led not by the government, universities or SOEs, but by individual entrepreneurs. This soon captured the attention of foreign venture capitalists, who started coming to China for investment opportunities. Eventually, venture capital proved to be a valuable instrument that further fueled domestic innovation and entrepreneurship.

The Place

Shenzhen was a municipality a bit smaller than Hong Kong in the 1990s. Back in 1950, when the new Republic was formed, Shenzhen existed as a fishing village. Hong Kong then was already a glistening international city, touted as the "Pearl of the Orient," under British colonial rule. Shenzhen's strategic importance was slowly revealed, as most refugees traveled across its border to Hong Kong. At the same time, capital, materials, technology and management resources from the outside world passed through Shenzhen and went north to China. It was a gateway to the gateway of the world.

As China adopted its reform and opening-up policy, Shenzhen was made a special economic zone (SEZ) in 1979, and as a result it enjoyed many policy measures to stimulate local development and attract foreign investment. Under this policy, many companies from Hong Kong, Taiwan and the Southeast Pacific overseas Chinese communities came through. Some gradually spread to inland provinces. Big name multinational corporations soon followed. Shenzhen acted as a major conduit in the process of national economic development. Many state-owned companies also set up operations or branch offices here to facilitate business expansion.

Gradually, people from various Chinese provinces moved to Shenzhen for access to better business opportunities, more job choices, a quasi-capitalistic lifestyle and a warmer climate. Many newcomers were elites and professionals who engaged in business, raised families and helped the city grow by taking advantage of the new socioeconomic conditions. Thus, Shenzhen was a typical immigrant city. Unlike other inland population centers, it was vibrant and dared to change.

Foreign investments brought not only capital and technology, but also new management practices. Through many joint development projects, efficiency and effectiveness became part of the growing mindset. One saw large banners hanging outside of skyscrapers under construction that said, "Time is Money; Efficiency is Life" ("時間就是金錢; 効率就是生命"). This publicly-proclaimed capitalistic attitude was considered antagonistic and even antisocial in many other cities that praised ideological and spiritual purity. However, it gradually became part of Shenzhen's culture.

With the convergence of businesses and people in and outside of China, Shenzhen increasingly became a place where international development and future technology trends were showcased. This highlighted the attractiveness of Shenzhen which further motivated the influx of people.

The Hong Kong Connection

Between 1981 and 1993, Shenzhen's economy grew extremely rapidly, with an average annual GDP growth rate of 40% while China's overall growth rate was just under 10%. This was the result of the concerted efforts of different government levels in conjunction with the city's SEZ status. With this status, many experimental and provisional economic practices were permitted. Whereas government resources focused on infrastructure building, private enterprises conducted business with efficiency. The close neighboring position with Hong Kong was a great help.

During this period, business collaboration between Hong Kong and the PRD, in general, and Shenzhen, in particular, could be described as a "Front Office and Back Factory" (前店後廠) operation model. Almost all manufacturing operations in the PRD had business offices or headquarters in Hong Kong to connect with their international customers. The financial and marketing services in Hong Kong, such as accounting, auditing, banking, insurance, legal and logistics management services, were readily available to them.

Many manufacturing operations in the PRD also used Hong Kong as a place to access new technology for process or product developments, and new methodology in management services. Basically, most of the Pearl River Delta's industries were of the "Three to One Mix" (三來一補) type and each side learned to work complementarily with the other. They eventually made use of the Hong Kong Stock Exchange (HKSE) to access private and institutional investors. Their stocks could be listed according to HKSE rules.

The close interactions between Hong Kong and the PRD proved to be mutually beneficial. They made the Greater Pearl River Delta one of the world's premier manufacturing centers and contributed to Shenzhen's rapid economic growth.

To sustain further growth, Shenzhen needed to strengthen its human resources and technology capabilities. Thus, investments in education and R&D became items of administrative priority.

In 1983, Shenzhen University was established as an institution of higher learning. To speed up the process, efforts were made to invite established institutions, such as the Chinese Academy of Sciences, the Harbin Institute of Technology, Tsinghua University and Peking University, to establish branch campuses in Shenzhen.

The return of Hong Kong to Chinese sovereignty in 1997 marked the beginning of a new chapter of the relationship between Hong Kong and Shenzhen. The central government established a policy goal to expedite collaboration between Hong Kong and the PRD. Being in a strategic location, Shenzhen was positioned to receive more education, technology and financial resources. The local leadership, both the party and administration, instituted many policy measures to maximize their effectiveness.

In 1999, the newly established the Hong Kong University of Science and Technology (HKUST) was invited to set up a joint R&D base in Nanshan District. The intent was to facilitate the technology development and knowledge transfer of the HKUST in the Chinese mainland by engaging Shenzhen as a strategic base station. As Vice President of R&D of the university, I took responsibility for its operation. The concept to establish a base in Shenzhen was soon adopted by all other major institutions of higher learning in Hong Kong. Meanwhile, the Southern University of Science and Technology (SUSTech) has expanded its curriculum, faculty and studentship to nurture the foundation of innovation and entrepreneurship.

The Action: Technology and Innovation Policy

After the Eleventh Five Year Plan (2005–2010), it was decided that innovation would be made a national focus. This motivated Shenzhen's government to formulate a pragmatic strategy. In July 2008, the Shenzhen City Fourth People's Congress established a general directive to promote technology and innovation activities as the vehicle for enhancing the city's competitiveness. The directive was further amended on December, 25 2013 to accelerate implementation. (深圳經濟特區科技創新促進條例, 2008 年 7 月 22 日通過; 2013 年 12 月 25日修正通過).

The directive specified the formation of an innovation system and emphasized building technology businesses by nurturing their fundamentals: human capital and technology resources. The 22 provisions of the directive focused on business as the leading entity of the innovation system and clearly stipulated the government's role in improving the system's environment factors.

It required the government to work out a strategic plan outlining its mission, objectives, resource investment, implementation, program review and evaluation. For example, to recruit experts from outside the country, the government would support qualified business entities by matching their R&D costs and providing the experts with housing, transportation and family support. Modelling after the State Ministry of Science and Technology's Thousand Experts Program, Shenzhen established the Peacock Program providing support to technology experts to attract them to come to the city to work for its technology business.

Shenzhen's rapid GDP growth between 2005 and 2015 is shown in Figure 17.

Shenzhen's R&D investments during this period were exemplary compared with those of neighboring cities and economic entities. Figure 18 shows a compilation of the R&D investments of Hong Kong, Shenzhen, Singapore and Taiwan, along with those of China and the U.S. as references.

Shenzhen remarkably stands out in comparison. Clearly, large scale R&D investment has enabled Shenzhen to carry out its various programs nurturing innovation and entrepreneurship, which in turn fueled the growth of its technology-based economy. This rapid growing process has aroused intense interest in many international studies.[83]

The Birth of an Entrepreneurial City

Shenzhen is at present distinguished by its entrepreneurial spirit and a megacity in its own right. It was home to many of China's successful

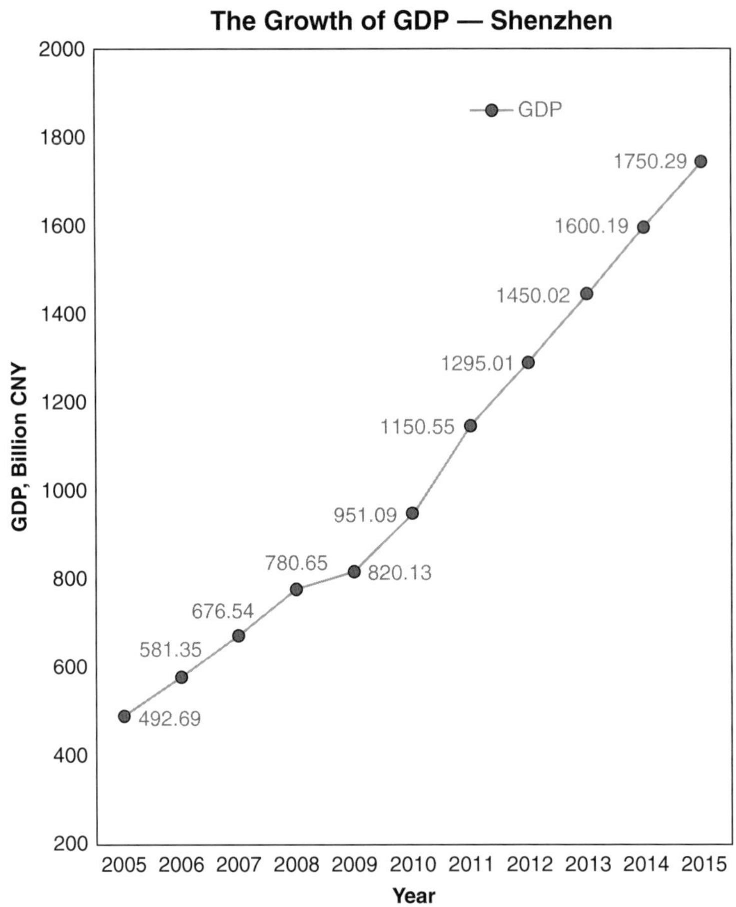

Figure 17. Shenzhen's GDP Growth (2005–2015)

high-tech companies, such as BYD, DJI, Huawei, Konka, Skyworth, Tencent, TP Link, and ZTE. Tencent Holdings Ltd. is now edging Facebook as the number one in chatroom membership worldwide (see Chapter 26). DJI and Huawei are world leaders in their field. Most of these companies are privately owned.

Other successful corporations headquartered in Shenzhen were China International Marine Containers, China Merchants Bank, Ping An Bank and Vanke. The Hon Hai Corporation of Taiwan also ran its largest manufacturing center in the SEZ through its wholly-owned subsidiary, Foxconn. It has become the major system integrator of the iPhone by Apple.

The Shenzhen Stock Exchange was established in 2010. Thus, parallel to establishing technical capability, Shenzhen built connections with the global marketplace and had access to global resources.

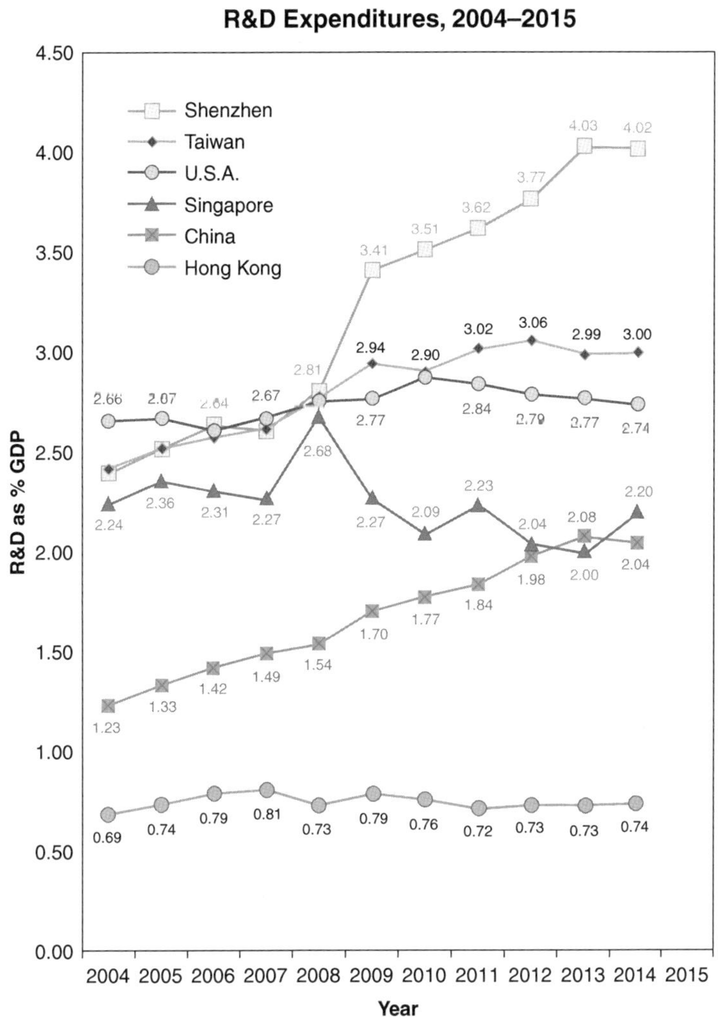

Figure 18. A Comparison of R&D Investment

Today the city Shenzhen is characterized by rising skyscrapers, tree-lined multi-lane highways, an airport serving national and international airlines, high-speed rails connecting Shanghai and Guangzhou, rapid metropolitan transit, public facilities, private housing, technology parks and university campuses. The fishing community of 332,000 inhabitants in the 1980s had exploded, becoming a modern metropolis of nearly 11 million. Through legislation, the city had also annexed some neighboring townships. As of 2016, it covered an area of approximately 2,000 square kilometers, an over six-fold expansion from 1980.

In 2016, Shenzhen's GDP reached CNY266 billion. It was economically the fourth largest city in China, ranking after Beijing, Shanghai and Guangzhou (hence, the group was generally referred to as "Bei Shang Guang Shen" or "北上廣深"). Its GDP growth rate was expected to be about 9%, the highest of the top four. Finally, its GDP per capita was US$25,000, the highest of all cities in China.

Looking Forward: Lok Ma Chau Loop

January 2017 started a new chapter of collaboration between Shenzhen and Hong Kong in innovation and entrepreneurship. Both cities' governments signed a memorandum of understanding to jointly develop the Lok Ma Chau Loop. The objective was to build 87 hectares of wetland into an innovation and technology zone jointly managed and used by both cities.[84]

The loop was in the bank of the Shenzhen River separating the two cities. It was created by Shenzhen's river-straightening effort in 1997. The State Council of China later decided that the loop fell within Hong Kong's borders. Negotiations regarding a mutually beneficial development of the loop has persisted for 20 years.

An agreement was finally reached in the beginning of 2017 to make it a technology zone as a home for technology start-ups from both cities. Hong Kong owned the loop and would pay for the environmental cleanup and infrastructure building. The zone would be jointly managed by both sides. Although the choice was made, many challenges remained.

The loop was four times the size of the Hong Kong Science and Technology Park and easily accessible from the mainland side. It appeared to be a wonderful opportunity to consolidate the factors of space, time, people and capital for both cities. A fundamentally important problem was promoting the free flow of people, technology and business without adversely affecting the judicial and political integrity of each. The zone is expected to nurture a truly harmonious habitat conducive to the growth of innovation and entrepreneurship, not to degenerate to become a simple real estate development.

Looking Forward: The Greater (G-HK-M) Bay Area

Another movement recently evolving is a proposal on the development of the (Guangdong-Hong Kong-Macau) Greater Bay Area for innovation.

The concept is to develop a megacity cluster to match the Silicon Valley–San Francisco Peninsula, the New York Metropolitan Area, and the Tokyo Bay Region. Nine cities in the Guangdong province bordering the South China Sea were included: Guangzhou, Shao-qing, Foshan, Zhongshan, Jianmen, Zhuhai, Huizhou, Shenzhen, Dong-guang. Added to these are Hong Kong and Macau. The group is alternatively referred to as 9+2. Indeed, the 9+2 Greater G-HK-M Bay Area consisted of the following characteristics in 2016:

- Land size — 56,000 km^2
- Population — 67 million
- GDP — US$1.3 trillion
- GDP per capital — US$20,000
- Harbour Capacity — 65 million TEU (number one worldwide)
- Air transportation — 1.76 billion passenger-trips
- Education — four of the top-100 world universities

This area stands to be in the same league as the Big 3 (Silicon Valley–San Francisco Peninsula, New York Metropolitan Area and Tokyo Bay Region). The Greater Bay Area already encompasses three development foci of the 13th Five Year Plan of the Guangdong province: Guangzhou-Nansha, Macau- Heng-qin, and Shenzhen Qian-hai. In view of the forecasted completion of the Hong Kong-Zhuhai-Macau Bridge by 2020, which will serve as a major conduit of the Bay area, this proposal has many obvious merits.[186]

The proposal is still in its conceptual phase and has received enthusiastic responses from the community especially those in the Shenzhen area. It is highly probable that Shenzhen as a world-class hub of innovation and entrepreneurship will be further highlighted in the future. However, many obstacles are expected to emerge. Not the least of these are regional balance within the province, freedom of movement of people, capital and information in the area, reconciliation of different culture and business norms, and others.

Perhaps a solution will be found on the basis of building mutual trust, soft power and the true entrepreneurial spirits of the people and the government. Hopefully, governments of all the cities involved will hone their entrepreneurial spirit in governance from the teachings of Laozi and Confucius (Chapters 14 and 15) to resolve these issues. The challenges are great but so will be the rewards.

Chapter 19

SINGAPORE: DEVELOPING A COUNTRY OF INNOVATION

In 1965, Singapore was a newly established independent republic with very little land, few natural resources, an unskilled workforce, a diverse ethnic composition and unfriendly neighbors. With political independence came economic uncertainty. This was the nascent state of affairs typical of the Third World. However, after 50 years of development, Singapore achieved first-world status in economics and wealth, and ranked in the upper echelon in terms of its national competitiveness. This could certainly be called a miracle. How was it made possible?

In the previous 25 years, I had visited Singapore many times, including a one-year sabbatical at the National University of Singapore (NUS). I was pleased to see a recent book edited by Hang, Seng and Thampuran on the Singapore research story, completed with significant details.[79] *In this chapter, I will attempt to provide some personal perspectives on how Singapore overcame monumental obstacles to achieve survival and prosperity. Let us first walk through the various phases of its development.*

Decades of Progress: Up the Value Chain

The 1960s

From the start of the republic, enriching society's livelihood was a major concern. Thus, Prime Minister Lee Kuan-yew established the Economic Development Board (EDB) to attract foreign investors and develop local industries. It soon became apparent that for labor-intensive industries to compete in the world marketplace, Singapore had to develop skilled manpower, build infrastructure, promote investment financing, raise industrial standards and conduct industrial R&D. Obviously, a multi-agency approach was needed. The EDB took the lead in establishing the Industrial Research Unit (IRU). The government concurrently established the Science Council to formulate national science and technology initiatives and enhance human resources. This was followed by the formation of the Ministry of Science and Technology in 1968.

Through the concerted efforts of these organizations, technical colleges, training groups, quality accreditations and technological conferences of various subjects were established. International expert panels guided many of these activities. In 1969, the IRU expanded to become the Singapore Institute of Standards and Industrial Research (SISIR). The "Made in Singapore" standard was introduced and the SISIR Marking Scheme was promoted.

The 1970s

In the 1970s, the government instituted programs to increase support for SMEs to enhance quality, productivity and new product development. Consequently, the industry gradually became more skill-intensive, enhancing the global status of Singaporean products and businesses.

At this time, a number of national surveys and studies highlighted the significance of developing education and technology for the future. The Ministry of Trade and Industry was formed. To secure social stability and nurture human resources, Lee took swift action to restructure the local institutions of higher learning. The NUS was chartered to be a comprehensive university, whereas the Nanyang Technological Institute, later renamed Nanyang Technology University (NTU), was formed to focus on science, engineering and technology. Multinational corporations (MNCs) showed growing interest in investing in Singapore for its strategic location for oceanic transport and its ability to add value to existing products. The total national R&D expenditure was approximately 0.05% of the GDP.

The 1980s

By the early 1980s, Singapore had progressed well enough to move into capital-intensive and technology industries. The Institute of System

Science, National Computer Board and R&D Assistance Scheme (RDAS) were all established in 1981. Under the auspices of the EDB, a number of major international technology companies, such as Glaxo, Philips Petroleum, Seagate, SGS and Sumitomo Chemicals, started to operate and build technical centers in Singapore. The Singapore Science Park was established to accommodate technology businesses. A biotechnology committee was also formed; it later recommended biotechnology as a part of the national economic development strategy.

With its newly gained expertise in managing industrial and science parks, Singapore became involved in establishing science parks and business and industrial parks outside of its own territory. This included development in Bangalore, India; Batam and Bintan, Indonesia; Ho Chi Min City, Vietnam; and Suzhou, China. This extended Singapore's influence beyond its national borders. With varying degrees of success, they added to Singapore's reputation of having expertise in managing multinational ventures.

A high-level elite group led by Tony Tan, Philip Yeo, Ming Keon Teo, C. C. Hang and others began to map a new national economic technology plan. The total R&D expenditure reached 0.85% of the GDP in 1990, with 54% coming from the private sector.

The 1990s

The National Science and Technology Board (NSTB) was established in 1991 to orchestrate strategies and provide guidance and support to national research programs. Many institutes were established to focus on selected fields of national interest, such as automation, data storage, microelectronics, materials sciences, molecular biology and software. All the institutes were under the NSTB's direct funding and guidance.

Additionally, the NUS and NTU engaged in establishing centers of excellence in fields that complemented their strengths. The SISIR, while expanding its international participation, merged with the Productivity Board to form the Productivity and Standards Board in 1996. This decade also witnessed many international technology corporations coming to Singapore to establish technical centers, some in partnership with the NUS, the NTU and local business groups. The list included Apple, ASM, BP, Compaq, Glaxo-Welcome, Hewlett-Packard, Lilly, Lucent, Motorola, Panasonic, SGS-Thomson (now STMicroelectronics), Shell and Siemens.

Several local technology companies, Biosensors, Creative Technology and Signet, were also establishing an impressive presence in the world marketplace.

The Defense Science Organization (DSO), which was dedicated to advanced defense-related technology research, was incorporated as a not-for-profit organization called the DSO National Laboratories.

Overall, the 1990s were a vibrant decade in technology development for Singapore. As a result, Singapore earned the reputation of a country capable of value creation in the knowledge economy age. The IMD repeatedly ranked Singapore in the top three for national competitiveness. The national R&D expenditure totaled 1.89% of the GDP in 2000, with 62% coming from the private sector.

The 2000s

In 2002, the NSTB was renamed the Agency for Science, Technology and Research (A*STAR). Singapore's economy had progressed up the value chain from labor intensive, skill intensive, value added to value creation. The new millennium was a time to focus on capacity creation. Therefore, A*STAR was designed to nurture human resources and promote technology development and innovation.

A huge laboratory complex, the Biopolis, was built to house most R&D activities in biomedical research. To complement the Biopolis, another complex for other science and engineering research, Fusionopolis, was established. Housing and living sites were later built to attract participating professionals, and the entire complex was grouped under the name One North. Likewise, the Productivity and Standards Board was renamed the Standards, Productivity, Research and Innovation Board or SPRING Singapore.

Several new research institutes were established dedicated to clinical sciences, bioengineering and nanotechnology, bioinformatics, genomics and medical biology. To develop and implement national research programs, the National Research Foundation (NRF) was established.

On the technology transfer side, A*STAR and SPRING Singapore started the Growing Enterprise with Technology Upgrading Program in 2003 to transfer technologies and researchers to SMEs. International bio-science companies responded enthusiastically to Singapore's initiatives:

Genentech, GlaxoSmithKline, Proctor and Gamble, Roche and Takeda started establishing factories and service laboratories. Furthermore, engagement with local businesses and universities increased. It was evident that Singapore's objective to capture the phenomenal potential of biological and medical sciences was on the right track.

The annual R&D expenditure in 2010 was approximately 2.10% of the GDP, with 60% coming from the private sector. This level of R&D effort was in the range of the OECD countries. Importantly, in 2005, the national R&D expenditure was already 2.36% of the GDP, which would have continued to increase if not for the global recession in 2008.

Key to Success

Singapore's success could be attributed to a government that led the establishment of an effective national innovation system. The system allowed scientific research, technology development and innovation to progress. The success factors included leadership, strategy, organization, commitment and systematic implementation.

These factors were closely interrelated. Leadership and vision were not empty words, but were backed by policy measures, adequate funding, organizational structure and concrete programs for implementation.

In the framework of the national innovation system (discussed in Chapter 6), Singapore's system was conscientiously built by a government that understood and was committed to both its fundamental and functional roles. The national innovation system was led by technology institutes and fully supported by the government.[129] The system was distinguished by its vigor and stability in organization and by its flexibility and agility in execution.

Singapore's innovation system was full and robust. At the top was the Economic Development Board (EDB), which was later reorganized into the NSTB. Also included was A*STAR, flanked by the NRF and the Research, Innovation and Enterprise Council (RIEC). On the organization level, the responsible players were the Biomedical Research Council (BMRC), the Science and Engineering Research Council (SERC), the SISIR and SPRING Singapore. On the execution level, the list consisted of various technical institutions: the DSI, the DSO, GinTec, the I2R, the IBN, the ICES, the IHPC, the IMCB, the IME, the IMRE, the ISS, the

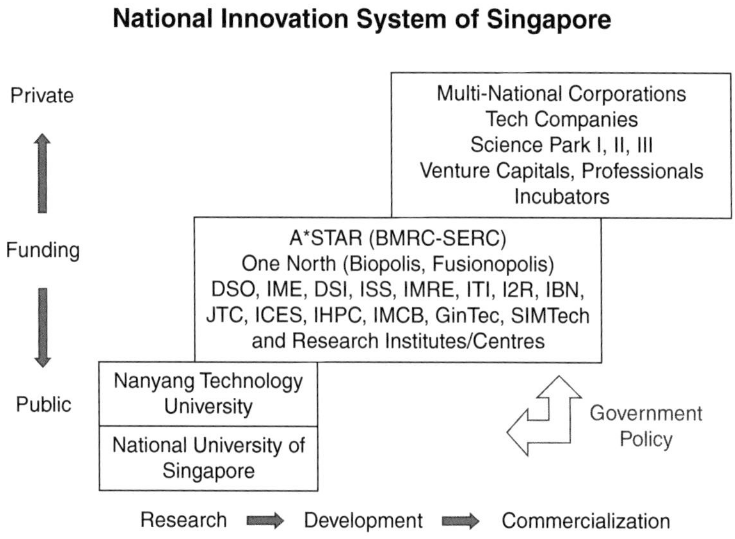

Figure 19. Singapore's Innovation System

ITI, the JTC, SIMTech and others. All these institutions were expected to swiftly adjust and adapt to the changing needs of the environment.

Figure 19 is a schematic diagram of the relationships between the players in Singapore's innovation system.

The full name of the institutions shown in Figure 19 are listed below:

DSO	Defense Science Organization
IME	Institute of Microelectronics
DSI	Data Storage Institute
ISS	Institute of System Science
IME	Institute of Materials Research and Engineering
ITI	Information Technology Institute
I2R	Institute for Infocomm Research
IBN	Institute of Bioengineering and Nanotechnology
JTC	Jurong Town Corporation
IHPC	Institute of High Performance Computing
IMCB	Institute of Molecular and Cell Biology
IMRE	Institute of Materials Research and Engineering
GinTec	Grumman International-Nanyang Technological Institute for CAD/CAM
SIMTech	Singapore Institute of Manufacturing Technology

In many other countries, setting up or closing down an institute or program could be painful and time consuming. This undoubtedly led to sustained mediocrity or perpetual bureaucracy. Somehow, Singapore was immune to this.

Human resources were another key feature. Singapore embarked on a very aggressive program to attract and recruit foreign talents of diverse expertise and backgrounds. Without the support of experienced scientists and engineers originally from China, India, Malaysia and Taiwan who were educated and trained in the U.S. or Europe, many R&D programs the Singaporean government boldly put forward would be very difficult to implement. This feature distinguished Singapore from some of its neighboring countries, which was prone to give preferential treatment to people based on religion and ethnicity rather than on professional competence and accomplishment.

Recognizing the importance of grooming indigenous human capital, the government initiated an ambitious long-term collaboration in education and research with MIT. While serving as a visiting professor at the NUS from 1996 to 1997, I was asked to help establish the collaboration between the NUS and MIT. After many discussions and visits led by Robert Brown, the Dean of MIT's School of Engineering, the program was established and quickly included the NTU. It ultimately became the landmark MIT-Singapore partnership which lasted for many years into the new millennium. The collaboration was a full-scale effort consisting of education, research and technical exchange components.

The Legacy: Soft Power

Pundits in Western political science tended to attribute Singapore's success to the strong-hand policy of its founding father, Lee Kuan-yew, who was educated at Cambridge University. Others accredited Lee's practice of Confucian doctrine to the national governance of Singapore. Lee's formal education in the U.K. is well known, but his contact with Confucian theory was relatively elusive. Thus, scholars were often puzzled at how Lee was connected to Confucius' teachings.

In my view, Singapore's success was rooted less in Lee's study of Confucian political theory but more in his behavior as a role model practicing simple Confucian values: loyalty to job, trust in people, learning and practicing.[130] These characteristics could be traced to his upbringing and reflected in his soft power leadership.

Over five decades, Lee demonstrated total dedication to the country he helped establish; built excellent relationships with colleagues, stakeholders and world leaders by gaining their trust; and stayed ahead of the globalization era through diligent learning and application. This behavior was natural to him because of his family values and his work with Chinese constituents ingrained in Confucian culture. Lee's leadership and role model demonstrated wide-ranging effects in Singapore.

The EDB/NSTB/A*STAR initiative was undoubtedly a major avenue by which Lee commanded to reach out to his people. The EDB/NSTB/A*STAR was not only the planner, but also the facilitator and coordinator. It was their job to lubricate the wheels of the innovation engine and help Lee achieve the goal of sustainable independence for Singapore.

As discussed in earlier chapters of this book, the innovation system of a nation was concerned with the formation of strategies, policies, implementation and modes of interaction between institutional players such as governments, universities, institutes and industries. A good, functioning system allowed entrepreneurs to pursue their goals more efficiently by turning research into value creation.

Generally, it was not difficult to establish such a system because for most countries the institutional players were already in place. The problem was in making the players interact harmoniously with mutual respect and trust.

Remarkably, the EDB/NSTB/A*STAR was able to collaborate and facilitate Singapore's national efforts, as the challenge was non-trivial. Singaporeans were generally recognized as dedicated, serious, hardworking and tenacious, while not easily compromised on their belief or position. A*STAR's role was to encourage people to think out of the box, find ways to reach out to others for compromise and collaboration to attain high-level national goals.[158]

Two types of collaboration were commonly needed. One was between government agencies (public to public) to minimize bureaucracy. The second was public-to-private to facilitate business development. The EDB/NSTB/A*STAR has played a key role in facilitating the institutional players and providing assistance to the entrepreneurs.

Thus, Singapore's innovation system was known for efficiency, fairness and transparency. This was conducive to building trust for MNCs, SMEs, shared service centers and individual entrepreneurs.

Over five decades, the country made every effort to learn the best practices from the outside world and spared little time applying them to fulfill its own objectives. Soft power characteristics were nurtured not just at the top, but also in the leadership at every level of the national eco-system.

The Way Forward

The world has witnessed rapid changes in technology and increased political and social turmoil in the new millennium. How will this affect the future of the Singaporean research community and its accomplishments in innovation and entrepreneurship? How will Singapore maintain its top level competitiveness in the world?

In the global arena of technology and innovation, the players include not only the traditionally advanced countries in North America, Western Europe and OECD, but also those in newly rising giants, such as the BRICS and ASEAN countries. Singapore had been active in almost every field of science and technology. Indeed, for a small nation with a population under six million, the ambition is admirable but the challenge seems overwhelming.

Limit Objectives on Selected Fields

The issue will lie in the assessment of challenges and in making choices. Similar to the Summer Olympics, it is hard for a city-state, however talented the people, to compete in all categories. Instead, it should aim for the gold in a few selected fields. Thus, it would appear prudent to review the long list of technologies currently being undertaken, to assess the challenges and set more realistic goals.

Collaboration and Partnership

The advent of internet technology in the globalization era can provide immense resources. Business now compete globally with no geographical boundaries. Thus, it might be possible for Singapore to excel in certain areas if partners could be identified and pursue with vigor.

Respect and Support for Individual Entrepreneurship

As with any human endeavor, bureaucracy and complacency will most definitively occur in time. This will hurt entrepreneurship. For Singapore, giving respect and providing support for private, individual entrepreneurship is an area that needs special attention. Since the government funded institutes are strong and situated in the mainstream, private individual entrepreneurs are placed in weaker positions, even sidelined. This is somewhat similar to individual entrepreneurs in big corporations. Individual entrepreneurs can be further disadvantaged if the overall habitat becomes more bureaucratic and thus less efficient. Experience in the world has shown that, more often than not, it was innovations from private individual entrepreneurs that made disruptive changes to technology and the society.

Build Synergy with Selected MNCs

In the past, Singapore welcomed all MNCs with open arms and the outcomes were generally mutually beneficial. However, the competitiveness of some MNCs have been eroded over time. Thus, Singapore would have to consider presently all available options and make careful choices. Furthermore, the recent uncertainties caused by Brexit and the changing U.S. foreign policy had complicated the progress of the globalization process. Singapore might have found that the global environment was not always conducive to its objectives as it used to be. Some MNCs may be pursuing options that could limit the reach of local entrepreneurship.

Learning from Legacy

It is hard to predict the future, but I continue to have high hopes for Singapore. I believe its future leadership will study the roots of its success and learn from its founding heroes: loyalty to jobs, build trust with people, learning new knowledge and practicing it to master new circumstances.

The recent dispute among Lee's surviving children on disposition of his house, a private estate with public historical significance came to light.[131] This might simply be a reflection of different values held and arguably a matter of trust in the family circle. Its resolution has, nevertheless, attracted the interest of those concerned with the future of Singapore on how the late leader's legacy will be interpreted and emulated.

Part Five
Role Models

Part Five discusses seven entrepreneurship companies at the initial phase of its establishment.

The start-up companies originated from the Greater China Region, including Taiwan, Hong Kong and the Chinese mainland, and all became global companies shortly afterward. They are all private entrepreneurs.

The business covered a wide range from personal products, industrial products, personal services, business services and electronic commerce. They dealt with different technology with diverse sophistication. They drew support from the various players of the national innovation system in varying degrees. Their accomplishments were exemplary in their own ways. I regard them as role models of entrepreneurship.

They are very different companies. However, in the initial phases of their establishment, some common characteristics can be found. I will focus on answering two questions.

— *What has driven the people to pursue entrepreneurship?*
— *What challenges have they encountered and what are the choices they made?*

It would be enlightening to examine the companies in terms of the spirit of entrepreneurship, including: people, teamwork, strategy, agility, quality, perseverance and trust.

Chapter 20

GIANT : CARBON FIBER BICYCLES

In the summer of 2010 during a visit to Taipei, I walk past a store at the corner of Xin-ye and Xin-hai Roads with a big shiny "GIANT BICYCLE" (捷安特) *sign. With great curiosity, I decide to walk inside and take a look. Displayed on the floor and the racks are bikes of various sizes, types and colors. Clerks are busily working with customers. Suddenly, I notice a very fancy gadget hanging from the ceiling as the feature presentation. Alas, it is more like a piece of art than a bicycle. The NT$350,000 (approximately US$11,500) price tag certainly indicates that. I gasp, "For that money, one can buy a Geely two-door sedan in Shanghai!" I double-check with the clerk, who responds, "Yes, sir. The tag is correct. And if you want one, we will put your name down on the waiting list. It may take two or three months for delivery."*

I recall a moment approximately 25 years earlier, when GIANT's President, K.B. Liu (劉金標)*, told me about the US$2,000 target price for a carbon fiber bike. I protested, "Who would ever want to spend that kind of money on a bike?" Politely, he said, "Dr. Lin, perhaps I should not count on you [to be among] my first group of customers!" I decided it might be fun to ask Liu what had transpired to put me further away from his product.*

The Carbon Fiber Bicycle Comes of Age

At 76, Liu looks exceptionally smart and youthful in his company uniform. After a lapse of almost 25 years, we shake hands enthusiastically and congratulate each other on this reunion. The meeting room at the GIANT Manufacturing Co. headquarters in Dai-jia (大甲), Taiwan, is almost the same save for the wall, which is adorned with many medals and trophies. Of those that are prominently displayed are National Quality Medals from different years dating back to the early 1990s for carbon fiber composite bicycles. Most have been the first in Taiwan.

I commend his athletic physique, which reveals no sign of the wining and dining that generally follows millionaires. He replies joyously and, aptly enough, that it is mainly attributable to the wonderful tool and, indeed, toy that has been his claim to fame — the carbon fiber bicycle. As his daily exercise, he likes to ride the bike to the office and, over many years, has participated in many "around the island" bike tournaments. As recently as 2009, he led a journey of bikers from Beijing to Shanghai over a distance of 1,668 kilometers. Quite obviously, he is the much-celebrated senior member of the group.

In many ways, Liu is the same person I knew nearly a quarter of a century before. He is courteous but straightforward, and never hesitates to speak his mind. I recall quite vividly our encounter in a public meeting in 1984. The subject was applying industrial technology to improve industrial competitiveness. He spoke forcefully on the plights of SMEs in Taiwan. He talked about the fierce competition in the U.S. market, although the average price of a typical bicycle was below US$100. He said he was squeezed from both ends by the glamorous but expensive bikes from Japan and the primitive but cheap bikes from India. The government and the ITRI, regretfully, did very little to help his industry.

As Director of the newly established Materials Research Laboratories (MRL) at the ITRI, I made it a habit during the mid-1980s to meet with industrialists around Taiwan. I wanted to gain more insight into the industrial problems and what the MRL could do for them. On those occasions, I discussed strengthening industrial competitiveness through the three basic elements of industrial products: design, materials and manufacturing. Analogous to the three legs of the ancient Chinese *ding* (古鼎), I called these the "three relevant technologies" of industrial products. They were of fundamental importance to any product, whether high- or low-tech. In this meeting with the sporting goods industry, Liu challenged me to talk specifics.

> Well, then, let us think about these elements for the bicycle. The design is always a frame with two wheels. There does not seem to be much room for change, right? The manufacturing process is complicated and you apparently know much more than I do. Besides, any suggestion to use

more machinery to replace labor would definitely add cost. What about material? Can you use new materials to make new bikes? Would the market accept a lighter bike — lighter but stronger?

Hearing this last question, Liu's eyes glistened. He had heard about the glass fiber technology that the MRL had been working on with the tennis racket industry. The industry was very pleased to find that users eagerly accepted these light-weight rackets. Players liked the new rackets, as they were easy to maneuver. Many athletes had actually improved their skills with them! However, it was unclear whether glass fiber could be strong enough for bicycles since more safety concerns were involved.

To this, I said I could give him carbon fiber, which had proved to be as strong as alloy steel and almost as light as a bird's feather. Carbon fiber had already been used in high-strength machine parts, such as helicopter propeller blades, which weighed only one sixth or one fifth of their metallic analogs. Furthermore, as plastics, they were immune to the corrosion that plagued metals.

There was not much more I needed to do to persuade Liu to move along this course of innovation. He was excited for an opportunity to achieve a higher level goal which I did not recognize at the time: his own brand. In the past, the company operated as an obscure original equipment manufacturer (OEM) for other brand name companies. The profit margins were extremely meager. This will all change as the use of new materials will be a game changer!

This was essentially how the development of carbon fiber bicycles in Taiwan started. However, the actual process was not as simple as expected.

Institute-Industry Collaboration

In the summer of 1985, GIANT Manufacturing Co. (巨大機械公司) and the MRL signed a contract to carry out an R&D project to jointly design and develop a bicycle frame made of carbon fiber and epoxy polymer. The MRL was to work on the underlining composite mechanics, formulate the material system, develop the basic fabrication methodologies and design tests to define the physical and performance properties (see Figure 20). On the other hand, GIANT was to work on manufacturing technologies in its factory, define a quality assurance system and conduct field tests, marketing promotions and of course, sales services.

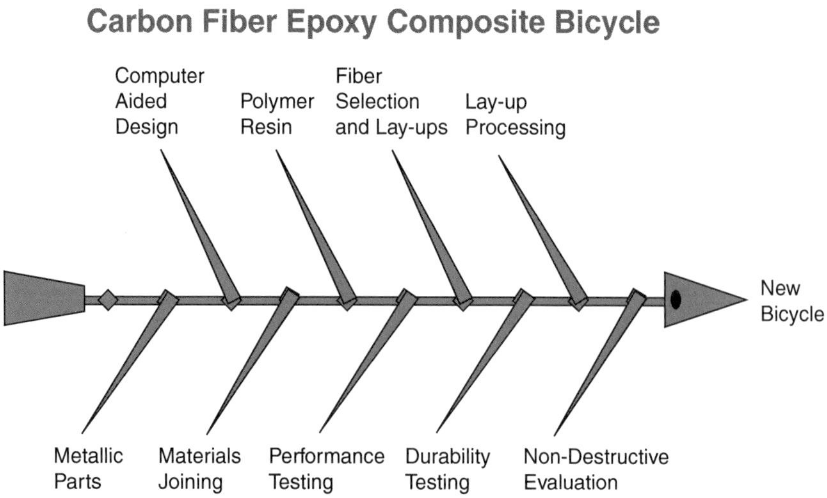

Figure 20. A Fishbone Sketch of Carbon Fiber Bicycle Technology

The MRL project was undertaken by the Composite Materials Group, which had been developing expertise on composite mechanics, polymer formulations and material processing with funding from the MOEA. GIANT funded the actual material and staff costs and related expenditures on the GIANT project. Ultimately, the MRL/ITRI would own any patents, but GIANT had preferential status for technology licenses for commercial use. K. B. Liu and his general manager, Tony Lo (羅祥安), spearheaded the project from GIANT's side. Notably, the Liu and Lo team leadership became a celebrated example of participative management in Taiwan's business community over the following three decades.

The MRL team was led by project manager Roger J. X. Liu (劉榮憲) and had approximately 10 staff members including: Reynold Liu (劉榮隆), C. P. Chan (陳中屏), C. L. Wu (吳宗麟) and D. M. Jiang (姜達銘) among others. It was composed of young graduates from local universities, such as NTHU, NTU and NCKU. Some of them had practical experience in the plastics and coating industries and were familiar with epoxy resins and glass fiber pre-pregs. However, none had ever worked with carbon fiber, which was a rarity in Taiwan due to its strategic importance at the time — it was an item under stringent Coordinating Committee for Multilateral Export Controls (CoCom) control.

I actually had some experience at Du Pont working with high-strength Kevlar® polyimide fiber, an important ingredient in high-performance composite systems such as Sikorsky helicopter, and other electronic printed circuit boards. The exposure had given me little else than the realization that we had a tough challenge ahead and needed expert assistance.

Globalization of Technology Sourcing

Luckily, a good friend, Dr. Steven W.L. Tsai (蔡維崙), research professor at Stanford University and a renowned scientist in composite mechanics well versed in product design, was traveling to Taiwan on vacation. Before joining Stanford, Steven was Chief Scientist at the Wright-Patterson Air Force Base. I asked Steven to come to the MRL to give the group of young scientists and engineers a cram course on the fundamentals of carbon fiber design and processing.

The group was thoroughly enlightened by Steven and worked diligently under his instruction. At the same time, Steven was quite moved by their diligence and insatiable desire for knowledge. He mentioned to me that seldom had he seen a group of such motivated students. In a relatively short time, our group had gained a good feel for formulating carbon fiber epoxy tubes and the property-performance relationship. At least they knew the subject matter well enough to carry out product development and evaluation on their own.

After approximately one year of hard work and trial and error, we were able to present to GIANT a number of material options for a carbon fiber epoxy bicycle frame. Each had gone through a battery of tests in our laboratory.

After hearing our report, Liu asked quietly, "How do I know that the bike will hold up with use?"

It was a legitimate question. Alloy steel and aluminum were common materials and had been used for many decades. People knew these materials well and were comfortable using them in new products. However, carbon fiber epoxy was like a the new kid on the block and seemed very strange to the community. Nobody knew it well, let alone had any confidence in it.

Globalization Outreach: The TNO

To address Liu's concern, we focused on assessing the durability of carbon fiber epoxy bicycles under realistic usage conditions. Obviously, we needed to establish special test equipment and methods, which would be costly and time-consuming to develop from scratch. Thus, we searched for any source that could assist us. The Netherlands came to our attention thanks to our early contacts with many global technology institutes.

Bicycling was a favorite sport in Holland. I had been told that there could be twice as many bicycles as the number of people in the Netherlands. Special bike trails were everywhere in the city and along roads and highways in the country. Naturally, the Dutch were very concerned about bicycle safety issues. The Netherlands Organization for Applied Scientific Research (TNO), a counterpart of the ITRI, had established a complete testing laboratory dedicated to bicycles. The laboratory could test bicycles with various loads, weights, positions and riding styles and under different surface and weather conditions. With this information, I visited the TNO to assess its methodologies and facilities. I promptly made arrangements for our prototypes to be tested there.

Based on the test results, we modified the formulations and lay-up configurations to improve the system. We repeated this until an optimum combination that yielded the highest durability was found. In fact, we ran the tests until many metallic parts were fractured or damaged, but the composite frame stood intact.

Liu was pleased with these data and directed his factory to begin manufacturing, with a product launch targeted for 1987. The company gave the new product lines a new brand name: 捷安特, which literally means "Fast, Safe, Distinctive", the phonetic translation of the company name GIANT in Chinese. The name was a stroke of genius. It summed up perfectly the salient features of the product and was easy to say and remember. The new product received public acclaim for its performance and was awarded the 1991 National Quality Gold Prize from the MOEA. The brand name remained stuck in the market ever since. I wish I had suggested that name but I did not!

The revival of the Taiwan bicycle industry was an interesting example of how new technology could transform old products and industries. Success of the GIANT carbon fiber bicycles has stimulated the confidence and interest of many SMEs of bicycle parts suppliers to upgrade their own products: gear chain, gear box, the wheels and even the design. Together this has created a new wave for the bicycle industry which was previously dubbed a sunset industry. SMEs slowly recognized that they could take advantage of high-tech which was, apparently, more than just microelectronics. There are other interested parties starting to market mountain bikes, or carbon fiber-epoxy fishing rods, and golf clubs, etc. This awareness gradually spread to SMEs of other industries, marking a revival of the traditional industries in Taiwan.

A Stunning Surprise

As I recount on that summer day of 2010 the major milestones of the development of the past 15 years, the conference room is inevitably filled with an air of satisfaction. Like a spoiled child expecting candy, I ask Liu whether that was also a happy moment for him. Characteristically, he responds, "No. This only brought me great anguish."

I am astonished. For a moment, I pull myself up, gasping for air and trying to prepare for further unexpected surprises. Most aptly, my research associate Jennifer Hou, who was sitting next to me, moved her chair to distract attention from my awkwardness.

He continues, "Happy times came a good deal later." He smiles, "This was natural, really. When your job was done, mine had just begun." And, "At that time, we were still not familiar with the technology. We were very busy trying to stay on top of many problems that sprung up on the production floor." It would cost him an additional 1,000 units worth of the carbon fiber bicycle, he said. Not a small sum indeed, for an SME at that time.

Although not directly involved with their work in the factory, we had expected the manufacturing process to be a complicated and strenuous one. Handling carbon fiber pre-pregs was entirely foreign to GIANT. Our formulation of the material system had required specific methods of laminating the fabric layers, the pre-pregs, followed by precise courses of thermal treatment to attain the desired physical properties.

This was hard. Doing it on a small scale, with scientists and engineers in the lab, was one thing; but doing it on a factory scale, with novice workers, was totally different. Any variation in the chemical composition, processing time, pressure, temperature or sequence would have been reflected in the final product. Above all, the factory was guided by the final product quality and production yield, which directly determined cost and profitability.

It later came to my attention that to improve its manufacturability, GIANT had instituted many quality engineering programs, such as

Six Sigma and LEAN technology. Although both sides had expected challenges in manufacturing, we had really underestimated the magnitude of such challenges. This was a textbook case illustrating the difficulty of commercializing scientific research.

Overcoming Roadblocks

The major difficulty was not in the material system or the manufacturing process. It occurred at the joints of the frame where the plastic met the metal. Materials joining had always been a difficult subject requiring a great deal of study of the specific materials system and methods being applied. Considerations of the MRL group based on general principles could be inadequate. But since no information from TNO test would signal any problem, the recommendation of the chemical formula and process were put forward as planned.

Any manufacturing defects due to minor variations in composition, processing and manufacturing conditions would be multiplied causing variations in product quality. As a technology conscious team, personnel on the floor and in the laboratory spared no efforts trying to resolve the problems they observed. But that was not enough.

While this was happening, some new bicycles were already out in the distribution channels. Alarmed that certain joint defects could occur with use over time, Liu voluntarily ordered an immediate recall of the products, well before any feedback was received from the marketplace. It was a costly decision. There were no reports of failure or any accidents with the new product. This might be just a precaution but obviously the right decision. Even I was left in the dark until this day.

Approximately 1,000 carbon fiber bicycles were recalled and buried underground as part of the foundation of the new factory laboratory building. According to Liu, "It was a painful decision to make." That I was ignorant of the pain he suffered 25 years ago leaves me with a very heavy heart.

I ask, "Why didn't you call me up at that time?" Seeing my anxiousness, Liu smiled. "Well, I thought it was time that we should learn to tackle it first, ourselves."

Although the MRL technical personnel frequently visited the GIANT factory, they were there to provide consultation and assistance and I was periodically briefed on the progress. When I became President of the ITRI in 1988, my personal attention to the project regretfully diffused gradually.

In retrospect, had I known the magnitude of the joint problems, I would have dealt with it promptly and sought outside help, if needed, to facilitate a solution. However, this information might then have leaked, thereby causing unnecessary publicity. Perhaps this was on Liu's mind as he weighed all the options.

Ultimately, by working on the problem alone, GIANT developed a better technical team and gained more insight that led to a solid proprietary position in carbon fiber bicycle manufacturing. This was what technology-based innovation was all about. This was his challenge and his choice!

Building Technological Capacity

In 1992, I noticed that GIANT had made remarkable progress in the development of a new product line of carbon fiber bicycles based on its own "One-Piece" process. The accomplishment further earned GIANT the 1994 National Quality Award. I was pleased that a technology recipient of the ITRI was able to continually improve its technology and strengthen its competitiveness.

GIANT was an exemplary ITRI technology partner, as it was able to develop its own technological know-how to compete in the international marketplace. With its good foundation in global marketing, distribution network and commitment to quality, the company quickly gained a worldwide reputation for its new products originating from modern materials technology. It earned many international awards and widespread recognition, further helping to expand its global operations. This is why I have valued technology transfer to the existing industry because they can rapidly carve a new position in the market. GIANT grew from a mid-sized Taiwanese manufacturer to a global company, with approximately 10 times the revenue in a few years. The carbon fiber bicycle was its centerpiece and was recognized as a Taiwan Brand product. I am pleased to have been associated with its early-stage development.

Responsibility of Sustainable Business

Listening to Liu's account of his decision to recall potential problematic products from the market, I am very much moved by his sense of responsibility to the customer and society. It reflects well on the value of the company and is an excellent lesson of business ethics.

I see an industrial business as a platform to provide products or services to its customers, which can be individuals or groups. These products or services must be of value to the customer and the society at large. Regardless of its nature, the business must uphold its word by selling products of promised quality, price and delivery. It can thereby generate confidence and trust among its customers and become a responsible citizen in the society it serves.

This is essentially what a Chinese proverb says: "信義為立業之本" or "Trust and justice are the foundation of business sustainability." GIANT's actions conformed to this principle.

It was no wonder that GIANT's products had generated customer acceptance worldwide. In 1985, GIANT's sales revenue was approximately US$100 million. In 2010, it exceeded US$600 million. The carbon fiber bicycle line accounted for approximately 2% of the total units sold, but nearly 20% of the revenue. Quality focus, customer orientation and social responsibility were not only good management practices, they were the foundation of globalization and thus made good business sense. These practices were behind GIANT's success over the past three decades.

Innovation with an Entrepreneurial Mindset

I ask Liu if he would still make the decision to launch carbon fiber bicycles today, knowing full well the difficulties that he would encounter. He responded, "Yes, most definitely." He stated that for a company to stay ahead in the game of global competition, it must embrace the challenges that come with continual innovation. He adds, "To be successful, a company will have to align with the consumers and understand their needs as time changes. You cannot stand still and be complacent while the world changes."

In the early 1990s, GIANT, with the carbon fiber products in its portfolio, was far ahead in the world marketplace. However, by then, many more players had entered the field, from Taiwan and mainland China alike. One needed to maintain the momentum of making progress in technology and innovation. During the last decades, GIANT embarked on new innovative areas, including the battery-assisted electric bicycle. Of course, fulfilling the requirement for a lightweight total system remained a primary focus. Liu had recognized that innovations in materials and energy technologies were very much in demand for the future.

Liu was an outspoken and unabashed lover of bicycling. He circled around Taiwan in bicycles many times to promote bicycling as a way of life. At his advanced age, this aroused a great deal of publicity. He was convinced that bicycling was a good exercise, non-polluting and enabled one to enjoy nature and appreciate life. It was a simple thing that everybody could do. Furthermore, it sustained energy efficiency and a low carbon footprint. Knowing that only 15% of people worldwide liked bicycling, he wanted to find the lost 85%, with almost missionary zeal. This "bicycling culture" was very much a part of his conviction for a green environment and harmonious society.

We shake hands most enthusiastically and bid each other farewell. As I walk out of the GIANT factory, I see workers busily working with many fancy and exquisite carbon fiber bicycle frames adorned with not only the GIANT label, but also other famous names. Obviously, GIANT is now the predominant manufacturer in its field. The workers are busy, but seem to be enjoying their work. They have contributed to the company's objective to achieve technology and marketplace leadership. They know well that the products they make will shine in many places with different faces.

Epilogue

At the end of 2016, GIANT Manufacturing Co. held a press conference in its flagship store in Taipei. K. B. Liu announced that he was retiring as Chairman of the company, which had been established 44 years ago. He was to be succeeded by S. J. Du, his niece, who had been the company's Chief Financial Officer for approximately 30 years. Tony Lo, Chief Executive Officer, was also retiring at the same time. He was to be succeeded by K. B. Liu's son Y. C. Liu, who had worked for the company for over 36 years. Y. C. Liu started as a quality engineer in the production line and was on the team that led the production of bicycles in Kun-shan, China. People in the industry had expected both retirements and successions.[133] Importantly, GIANT had been recognized as the number one bicycle manufacturer in the world under the stable leadership of Liu and Lo. By 2016, foreign investors held approximately 54% of the company equity. The smooth transfer of the leadership to its next generation was also exemplary.

Chapter 21

EPI-STAR OPTOELECTRONICS: HBLED

In a meeting at the ITRI in the early 1980s, President H. C. Fang asks me to look at an experimental facility located near the library of the Union Chemical Laboratories (UCL). He wants to know what I think of it. He asks, "Should we keep it or give it away? You see, the facility houses an old chemical reactor that makes certain materials that no one has found use for. But it processes many toxic chemicals and is considered a nuisance for hygiene and a source of potential hazard to its neighbors. Besides, the UCL finds it expensive to maintain."

At this time, I have returned to Du Pont after a one-year sabbatical leave at the National Tsing-hua University in Hsinchu. However, a few times a year, I travel to Taiwan as an advisor. Fang thinks that I, being a material scientist, should know more about this topic, if I am to be of any good.

After visiting the laboratory, I realize that the reactor is a crystal puller, or grower, that makes compound semiconductor single crystals, such as gallium phosphite (GaP) and gallium arsenide (GaAs) using the Czochralsky method. Although having no personal experience with it, I know materials consisting of the III-V family of elements from the periodic table have the ability to emit light. Herein lies the foundation of optoelectronics, the much-touted fifth-generation computer. Thus, I recommend keeping it. I argue that even if it is of no obvious use now, it can serve as a training platform later for the future generation of optoelectronic technology.

Although Fang is skeptical on the fifth-generation computer, he thinks my point about training is a plausible consideration. Therefore, he decides to save the facility.

Birth of Optoelectronics Industry in Taiwan

Fang would be proven correct on both counts, unwittingly at the time. The fifth-generation computer never materialized as expected. But the scientists and engineers the facility trained became trailblazers of optoelectronic technology, contributing to a booming light-emitting diode (LED) industry in Taiwan approximately 15 years later.

This industry helped to add over US$4.55 billion to the annual gross national product in Taiwan in the year 2010, making Taiwan the number one LED manufacturer in the the world. The revenue rapidly grew to about US$18 billion in 2015.[134] With competition growing from Korea, Japan and especially China, Taiwan still ranks among the top three today. It is a field ITRI proved to play an important catalytic role and has assisted Taiwan to strive in the era of optoelectronics.

Debut at the Materials Research Laboratories

In 1983, I was appointed Director General of the newly established Materials Research Laboratories (MRL) at the ITRI. Fang thought it was natural for me to take over the compound semiconductor crystal growth laboratory at the UCL, a responsibility that I could not refuse. Thus, I established optoelectronics as one of the MRL's strategic focus.

The importance of III-V compound semiconductor materials was evidenced. Developments in the U.S. and Japan have shown that formulated as LED, it is a light source without light bulbs. The material in its solid state emits light under small electrical excitation. It offers many advantages over the incandescent light bulb with its compact size, long lifespan, illuminance, durability, safety, low maintenance and little pollution. For application as display, it can be used in small devices as well as large panels, such as for athletic fields. For application as space illumination, it can be used in private house, office tower, factory, airport and highways. Overall, it offers opportunities for many new industries in Taiwan.

Thus MRL started an active recruitment drive to form a research group on optoelectronic materials. At this point, the leading universities in Taiwan: National Taiwan University (NTU), National Tsing-hua University (NTHU), National Jiao-tong University (NJTU) and National Cheng-kung University (NCKU) have established respectable graduate programs in science and engineering. MRL was a top choice for the graduates. In a few months we were able to put together an outstanding group of young scientists. In all, by 1988, the group has grown to about 30 professionals, most with Master's degrees.

B.J. Lee (李秉傑) was in the cohort of young scientists who came to the MRL in 1985. As a fresh chemical engineering graduate from NTHU with a doctoral degree, Lee came highly recommended. Indeed very quickly his technical expertise and congenial personality made him a key figure among the group members. At that time, the majority of the group consisted of electrical and electronic engineering majors who had a natural phobia of chemical formulae. Having specialized in chemistry, Lee's role was quickly established in the group. He understood very well the periodic table of chemical elements and how compound semiconductors can be made more lively and versatile in comparison with silicon. He was to stay with the ITRI for more than 10 years until 1996, when he ventured into an entrepreneurial career.

Building Technology Infrastructure

After settling down at the MRL, I invited a number of distinguished scientists to be advisors of the laboratory. Among them were Harry Gatos of MIT, Gilbert Chin and Alfred Cho, both of AT&T Bell Laboratories in Murray Hill, New Jersey. Both Chin and Cho are of Chinese origin and were outstanding researchers in optoelectronics with international repute. Gilbert was Research Manager for a large group at Bell Labs. Alfred had made many fundamental discoveries in semiconductor physics and was widely regarded as the "father" of molecular beam epitaxy (MBE), a powerful method for studying molecular surfaces and thin films.

All three advisors suggested that the MRL should move quickly into the field of LEDs and explore some novel applications with the industry, display or lighting or both. Alfred Cho further suggested that we follow along the value chain, from crystal growth and epitaxy synthesis to selected end-use applications.

Very quickly, we decided that the MRL should move beyond single crystal growth into the area of epitaxy. Thus we needed to purchase a key instrument: a molecular organic chemical vapor deposition (MOCVD) reactor for its application versatility.

In the mid-1980s, an MOCVD reactor cost approximately US$1 million and was considered a big-ticket item. The MBE reactor would be more. This was why they were often dubbed the "mega buck evaporators." I estimated that through applying for research grants from the MOEA, we could fund the purchase in due course. In any case, we needed to prepare the lab for the instrument so that it would function quickly and efficiently. Thus, B.J. Lee, as the group's lead chemist, took up an assignment to work on the processing details of the MOCVD. I also requested the assistance of the Center for Industrial Safety and Occupation Health-ITRI (Ada Wen-Shung Ma, Director) to visit Bell Labs for advice on setting up the MOCVD laboratory which involves special controls of very hazardous chemicals. Being champion of the project, Lee took up the assignment quite happily. He did not know that this instrument would become his toy for life or for at least the next two decades.

Market-oriented Mindset

The overall optoelectronic materials project was supervised by Coy-Ing Nee from 1985 onward. Nee had established close contacts in the optoelectronics industry in Taiwan, which was just budding and consisted mainly of packagers. Except for TI which has a wholly-owned subsidiary and a local packaging specialist, Lite-On Ltd, most companies were very small working on simple product packaging. Noting that these packagers were often led to cut-throat price competitions with one another as maneuvered by foreign buyers of LEDs, Nee organized a technology-industry liaison committee as a forum of technical exchange.

As a major activity, the committee sought to collect all market and technical information related to product orders, make it anomalous, normalize it to certain common standards and distribute it to member companies for reference. This put the members in a better position to appraise market trends and develop their own plans. This small effort was appreciated by the committee members, which were initially no more than a dozen. But the membership grew quickly. It prompted MRL to play the role of an honest broker for the industry in subsequent years. With industrial confidence and continued participation, the MRL was also able to better align its technology programs with market needs.

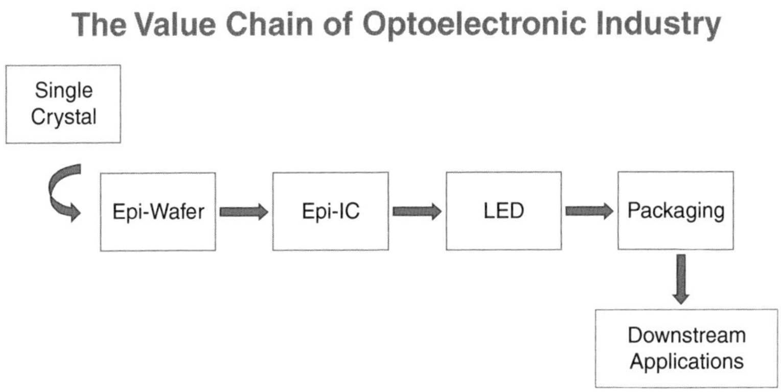

Figure 21. The Value Chain of Optoelectronic Materials and Technology

As the Liaison Committee continued, it evolved into an industry and trade association and played significant leadership in furthering the interests of the industrial community. This model of operation was repeated in other technology areas. Non-government Organizations like this one for optoelectronic industry that ITRI initiated, although relatively elusive at the time, were to become important players in Taiwan's future economic development.

Shortly after, I persuaded Ming Hseung Lin (林敏雄), a former colleague at NTHU who was the head of the optoelectronic engineering department to join the MRL. Under Lin's leadership, the scope and depth of the group expanded further. It became clear that by taking a system approach in integrating materials science and other application technologies, the versatility and potential of compound semiconductors could be maximized. This view became the MRL's shared vision.

The group was soon elevated in its organizational status to become the Center of Optoelectronics of the ITRI, under President Morris Chang. When I became President of the ITRI in 1988, the center grew further to become the OES Laboratories, with the same status as the MRL, UCL and ERSO. When I left the ITRI in 1994 after my two terms of presidency ended, Chintay Shih (史欽泰) succeeded me as President and Lin soon became the Executive Vice President. Eric Lean (林耕華), an ITRI fellow who was formerly a senior scientist at IBM, took the helm at OES Laboratories. All the while, B.J. Lee's career grew along with OES Laboratories.

Changing Roles

In-depth research in optoelectronics technology and broad based contacts with the industry had given the group members, and B.J. Lee especially, a clear understanding of the challenge and opportunity onward. Observing that most Taiwan companies were working on downstream and low margin business, Lee identified a niche in the upstream epi wafer supply.

In 1996, B. J. Lee and two other young colleagues decided to establish the company Epi-star Opto-Electonic Co., Ltd., dedicated to manufacturing of expitaxy wafers. One of the two, Ming-jun Chou (周銘俊) later became President of Epi-Star while B.J. Lee was Chairman and CEO. The other, Jacob Chang-lin Tarn (譚昌琳) was later to become president of an Epi-Star subsidiary Jin-Yi (晶誼) optoelectonics.

In the 10 years since 1986, the optoelectronics industry in Taiwan had grown rapidly to nearly 1,000 companies. However, it still consisted mainly of electronic packagers pursuing different end-use applications. There was a strong demand for up-stream supplies, including epi-wafers and epi-chips. Figure 21 shows the value chain relationship of the industry.

The single crystal and epi-wafer are the up-stream. Epi-IC where the wafers are processed through expitaxial diffusion, metal evaporation, masking, etching, heat treatment followed by creation of grain electrodes to form LED chips are the mid-stream. Packaging where the LED grains become discrete components or incorporated with other electronic and electrical components as modules are the downstreams.

At that time in Taiwan, there were many LED packages for various industrial applications. But chip suppliers were only a few and, all epi-wafers were imported. Therefore, the industry welcomed the arrival of an epi-wafer manufacturer in the local community. Indeed, a stable supply of good quality epi-wafers/epi-IC would enhance the overall competitiveness of the optoelectronics industry. The MOEA was very much in favor of building up an infrastructure for the expansion of the Taiwan optoelectronic industry.

Assessing the business environment, B.J. Lee felt that the time was right to take action. In March 1996, the Epi-star Corporation was formed under the auspices of the Industrial Technology Investment Corporation (ITIC), a wholly-owned subsidiary of the ITRI established in the mid-1980s to foster the commercialization of technology. Among many successful ventures of the ITIC was the establishment of the UMC in 1986 and the Taiwan Mask Corporation. Both were the first of their kind in Taiwan. I served as Chairman of the ITIC briefly in 1987. After I became President of the ITRI, the ITIC was led by Chi Mo Wang and Shirley S.L. Chang, both of whom worked full-time for the company.

Epi-star was the first company in the up-stream of the optoelectronics value chain and the first one in that industry to spin-off from the ITRI. Thus, the ITIC helped organize the new venture. It consisted of seven investors/partners representing the members of the value chain, the venture capitalists and the founding team. The total investment made was NT$320 million. Approximately 5% was cash contribution from the founding team.

A technology license was taken from the ITRI. It included intellectual property (IP) rights and a technology collaboration agreement to provide the new company with technical support over a three-year period. This package was valued at approximately NT$70 million, nearly 20% of the total investment.

Challenge of Entrepreneurship

When meeting with B.J. Lee again in July 2011, I ask him pointedly, "In retrospect, as an entrepreneur, how do you assess the value of this license package. Was it worthy of the money?" His answer is positive and concise: "Definitely. With formal technology transfer and the promised support of the ITRI, the market viewed Epi-star with confidence, which was a big help." He adds, "We could also count on the ITRI's assistance with its vast resources in technology." The ITIC's efforts to explore and expand market and business contacts were also helpful to the new company.

However, entrepreneurial life was a culture shock. First, they were no longer ITRI employees with the trappings of stable salaries and employee benefits. They were on their own. As the umbilical cord had been severed, feelings of freedom and independence were accompanied by uncertainty and insecurity. Second, they had all become debtors. The debt burden on each one's back was nearly 200 times their monthly salaries. Lee commented, "We will have to work as volunteers for the rest of our lives if we default on those debts." Thus, all of the teammates knew that they had to work very hard and very smart to succeed. There was no alternative.

Changing Business Strategy

They soon found that the original business plan had serious flaws. The plan called for the production of epi-wafers as supplier to makers of epi-chips. This was logical, but they had only two customers in the business of manufacturing epi-chips in Taiwan. Actually, at that time, there were only five such manufacturers around the world. Furthermore, both its Taiwan customers were investment partners and members of Epi-star's board. Any attempts by Epi-star to establish direct contacts down the value chain to explore business opportunities with the packagers would create conflicts for the investment partners. This placed serious restrictions on market expansion.

Eventually, against much resistance, Epi-star was able to reorganize the investment structure. These two investors were basically bought out. This has enabled the company to move the business forward along the value chain. As a supplier of customer epi-chips, the company gained direct access to over 2,000 customers.

The balance sheets for the first two years glared with red, reflecting a long initial phase of product commercialization. The third year showed a profit, but the fourth experienced loss and the inventory climbed rapidly. Then, the IT bubble broke. This was shocking, but luckily not devastating. The industrial demand was strong and Epi-star was able to endure a short period of downward market. The company went public on the Taipei Stock Exchange Technology Board in 2001. This started a new phase for the company.

It was a critical juncture for Epi-star to recognize the reality of the marketing channel and implement a new business strategy accordingly. This was the challenge of entrepreneurship in a market economy. Change was necessary for survival. In other words, private entrepreneurs are extremely hungry, and equally persistent in keeping their head above the water. This was quite unlike working for an SOE or big corporation where they are to stick to the course set by the upper management. In the end, there could always be a Plan B. But for an entrepreneur the choice is limited and one has to make the choice swiftly.

Making a Leader in HBLED

This was basically a change in the business model for the new company. They were adjusting to a model of forward integration to the mid-stream. This would help them to access a much larger market and achieve economy of scale. The transformation from an epi-wafer company into an epi-chip company was proven successful. Nearly 98% of the sales revenue came from the chips. Blue LED (InGaN) accounted for 75% of the chips and red and yellow LEDs (AlGaInP) accounted for the remaining 25%.

In 2010, the global sales revenue of High Brightness LED (HBLED) totaled approximately US$5.45 billion. Japan was the largest producer with 35%. It was followed by Taiwan and Korea, each with 19%; Europe with 15%; the U.S. with 9%; and China with 3%. Epi-star was the largest producer in Taiwan with 42% of Taiwan's HBLED sales revenue. It was nearly 2.5 times larger than the next one in line, the OIC, one of Epi-star's investment partners.

Epi-star provided HBLEDs of the full color spectrum, from infrared and visible to ultraviolet. This was attributed to its technological expertise in making wafer chips of multiple elements: GaP, GaAsP, AlGaAS, InGaN, AlGaInP and AlGaInN. It was also capable of making wafers and chips for optical sensors and other specialty electronic devices.

In 2010, Epi-star sales in LED was NT$12.7 billion and became the number one LED manufacture company in Taiwan. In 2016, the sales in LED grew to NT$25.5 billion and remained number one. It also held the first place in market share in epi-IC worldwide.

Technology Intensive Manufacturing

I was quite astonished by Lee's description of the manufacturing process. He said that there were over 200 units of MOCVD in use on the production line in his factory. How incredible!

I recalled 20 years before when Alfred Cho explained to me the high level of quality control and assurance needed for the composition and morphology of the chips. I said, "This would appear to require an MOCVD for quality control in each factory," which, I added, "would be impossible not only for the high capital cost, but also the high level of technical skills needed to run the instrument." He said, "One day, Otto, people will learn to use MOCVD or MBE as production equipment on the processing floor." I nodded courteously, but could not hide my skepticism.

I told Lee this story to show my admiration of a true visionary. How lucky we were to have been guided by such pioneers as Alfred Cho. Twenty years ago, we scratched our heads to find a way to fund the first

MOCVD for Taiwan. Now there are over 500 units in total, mostly in the industry. This is an indication of the changing technology content in Taiwan's manufacturing industry.

Epi-star's technology and innovation capability did not stop at the original technology license agreement with the ITRI. The company had invested quite substantially in R&D. Of the total 3,168 employees in 2010, 960 were designated R&D and technical. Approximately 3.6% of the sales revenue was earmarked for R&D expenditures. At present, it held approximately 3,000 patents.

Additionally, its technological capacity grew remarkably through continuous technical collaboration with academic institutions, such as the NTHU, NJTU and NTU. In 2000, the company entered into another collaborative project with the ITRI to develop AlGaInN, which helped to extend the company's capability into the blue and blue-green regions. It has also excelled in high illuminance and high power applications. In many ways, Epi-star was a good example of how innovation and entrepreneurship occurred in Taiwan.

Competition in the Globalization Era

Epi-star was globally recognized for its products. Most of the world's opto-electronics packaging activities were in the Asia Pacific region since the early 1990s. However, their client company's headquarters and users were international. Over 60% of the customers had headquarters or centers of operation located in mainland China.

During the last decade, China had aimed to develop its own capability in optoelectronic technology and manufacturing and has made great strides in the industry. Under the support of government policy, many optoelectronics companies were born, especially in areas where Taiwan's technology companies operated. For example, two major Chinese companies, San'an Optoelectronics Co., Ltd (三安) and Qian Zhao Electronics, Inc. (乾照), were located in Xiamen. Both these companies offered very aggressive salary packages to skilled professionals in the field. An estimate of over 100 engineers experienced in LED technology and business from Taiwan migrated in 2009 and 2010. China's LED industry grew extremely rapidly in the last decade. In terms of capacity, it stands as number one worldwide; but in terms of revenue, it may be number two.

Industrial leaders in Taiwan took the migration of technical people to mainland Chinese competitors as a problem of deepest concern. They considered this raid of technical capacity as an infringement of intellectual property right.

Large-scale mobility of professionals was hardly a unique problem in modern day China. During the late 1980s, I encountered a similar situation in Hsinchu. With the booming electronics industry, professional staff members at the ITRI had become easy targets for headhunters. Any mid-level engineer with 5–10 years of experience often became the target. In addition to attractive salary packages, new companies commonly offered lucrative bonus and stock options. At its peak of flux, the ERSO, the ITRI lab specialized in the IT field experienced a turnover of nearly 35% in one year.

The crux of this problem was the dislocation of needs between employees and their employers. If employees found that their organizations could not address their needs and aspirations, separation was just a matter of time. Therefore, a good organization should understand the motivation of its people and communicate with them frequently and promptly for mutual benefits.

In the case of ITRI, major organizational efforts were launched between 1989 and 1991. The program included reviews of the compensation package, restructuring management systems, revamping retirement schemes, improving the work environment, repositioning organizational objectives and establishing an organizational culture of mission and pride.[62] Externally, we discussed with the industry for a better technology transfer mechanism, such as the voluntary transfer of professionals under mutual agreements. All of these actions improved the ITRI, increased client satisfaction, made employees happier and, more importantly, assured those who chose to stay. The professionals were confident that the organization had kept their best interests and career goals in mind. Those who left knew that they had a clear responsibility to respect IP rights law and abide by a professional code of conduct. They were not only answerable to the organization, but also to their fellow workers and comrades.

However, in the context of the Chinese mainland, the situation was much more complicated. There appeared to be few, definitely inadequate, laws or codes binding people and companies on IP protection. Until professionals and corporations could nurture a common value of mutual respect, the situation would continue and be detrimental to the overall process of innovation. This was indeed a challenge for Epi-star and other companies in similar situations.

There was another issue related to technology and product standards. The scientific community advocated the establishment of common standards for future advancement of science and technology. However, many industrial groups and national governments had different views and priorities. It was no surprise that technology and product standards had been used either as an instrument or a barrier to globalization by various countries, which presented an area of concern.

Internationally, epi-wafer and epi-chip specialists would team up with large conglomerates for vertical integration as well as captive applications of optoelectronics. Examples included Luminus collaborating with General Electric to form GeLumed, and, Samsung with Philips for another venture. This forward integration became a formidable commercial barrier to young companies such as Epi-star, and, indeed, many budding companies in China and other developing economies.

Finally, in 2012, Epi-star formed a joint venture with the Advanced Photoelectronic Technology Co., Ltd (APT Corporation, headquartered in Nansha, Guangzhou) to establish a strategic partnership. The APT Corporation was an entrepreneurship company of the HKUST established by faculty and students when I served the university as Vice President of R&D. It specialized in energy efficient packaging and started its operations at the Nansha IT Park (Guangzhou) in 2005. The APT Corporation was publically traded in 2015 in Beijing's new Third Board of Growth Equity Exchange. I was pleased to see technology companies on both sides of the Taiwan Strait find ways to cooperate to increase their mutual competitiveness in the world market.

Interestingly, ITRI researchers founded Epi-star. In the last 15 years, it spawned many other companies pursuing LED technology products to operate in Taiwan and China. This illustrated the process of technology diffusion through innovation and entrepreneurship. In the globalization era, the process spread across national boundaries as dictated more by market forces than any ideological doctrine.

Chapter 22

ARDENTEC CORPORATION: IC TESTING

Ardentec was founded by four former ITRI engineers who were key members of the Vanguard International Semiconductor Manufacturing Company (VIS) established in 1994. VIS was established to commercialize the state-of-art dynamic random-access memory (DRAM) technology developed by the ITRI. As a start-up, the VIS turned in substantial profits within only two years of operation. This led to a frenzy of new investments in DRAM manufacturing in Taiwan. However, in 1997, the world's electronics market started to show a downward trend due to a supply-demand imbalance. Many international players started to bow out and the VIS's corporate management decided to follow suit, leaving the field of DRAM for the dominance of Samsung from South Korea.

Disheartened by the decision, leading engineers decided to go into entrepreneurship to pursue a different niche in the microelectronics value chain. They have the vision to carve out a new strategic position in the existing integrated circuit (IC) fabrication process to make it more efficient and economically sustainable.

Thus Ardentec was born in Hsinchu in 1999. It provided specific testing services for IDMs and IC foundries. Within 5 years, it achieved a compound annual growth rate (CAGR) of nearly 50%. In less than 10 years, it became a leading firm in its field with global acclaim.

Entrepreneurship with a Purpose

In 1988, while I served as President of the ITRI, the institute was planning to undertake a major project to develop the manufacturing technology of submicron very-large-scale integrated circuits (VLSI). The specific goal was to master all the technologies needed for the fabrication of a 0.5-micron DRAM on a 6-inch wafer. This would enable Taiwan to compete as a world-class player in state-of-the-art microelectronics.

The project was to be the largest R&D effort ever undertaken by the ITRI. It had a forecasted budget of approximately US$200 million, which was an astronomical figure for a R&D project at that time in Taiwan. Planning for the project had gone through many levels of expert review and had received government approval in principle for funding. In parallel with the planning undertaken by the ERSO/ITRI, a committee was established to conduct an international search for the champion to lead the project.

The search committee identified C.Y. Lu (盧志遠) as the prime candidate, but was unsure whether he would accept. At the time, Lu was a senior member of the Bell Laboratories in Pennsylvania. He was highly regarded for his technological accomplishments in VLSI. Within Bell Labs, he was considered as a rising star. Coming to Taiwan would require him to take a substantial pay cut, live in a relatively underdeveloped environment and face an uncertain future. However, the mission was exalted, the challenge was great and the opportunity was extraordinary. I could feel his motivation to take on the challenge, but also understood his hesitation.

On the last day of his visit, I took Lu to a technical conference in my car. After a review of the status, he asked, "Do you really want to do this project knowing its scale and difficulty? Are you prepared to carry it all the way?" I looked at him in the eyes and said, "We are all ready for the challenge. We will kick it off when you come."

Very shortly after, Lu reported to the ITRI for duty and the project was set in motion. Approximately 300 engineers, bright and fully charged for the mission, were recruited to join the team. As soon as he came on board, Lu recommended changing the 6-inch wafer target to 8-inch, which was to become the standard of the next generation technology. This was a very bold and futuristic strategy, and would create additional technical and administrative difficulties. However, we decided to make this choice and faced the challenges as they arrived.

In less than five years, by mid-1993, the project was successfully completed. Taiwan's ability to manufacture a 0.5-micron line 16 M DRAM was startling news to the global microelectronics community.

Under close government supervision, the ITRI started the technology transfer process by open tender, inviting the private sector to pursue commercialization. Finally, an industrial consortium was organized for interested parties with the TSMC as lead investor. With this, the VIS was established. At this point, I had completed my two terms of presidency and left ITRI. But I was pleased to learn that Lu was selected as Vice President and later appointed President.

The first two years of business operations were very successful and gave Lu and all the ITRI team members great satisfaction. Needless to say, the subsequent decision to pull out from the DRAM market in 1999 was a disappointment to many ITRI alumni who had started at the very beginning of the project. C. Y. Lu, together with Chi Ming Chang (張季明), Xiao Long Chin (秦曉隆) and Elvis Deng (鄧茂松) decided it was time to start another venture on their own to pursue their vision of boosting the microelectronics capability of Taiwan. They managed to raise NT$2 billion from local industries, venture capital and the ITIC/ITRI. Neither the TSMC nor the VIS had participated in this venture.

Identifying a Strategic Opportunity

The new company, Ardentec, was positioned at the center of the semiconductor manufacturing value chain. It linked the "front end" of IC design and fabrication with the "back end" of IC packaging and product testing. The schematic diagram shown in Figure 22 illustrates the

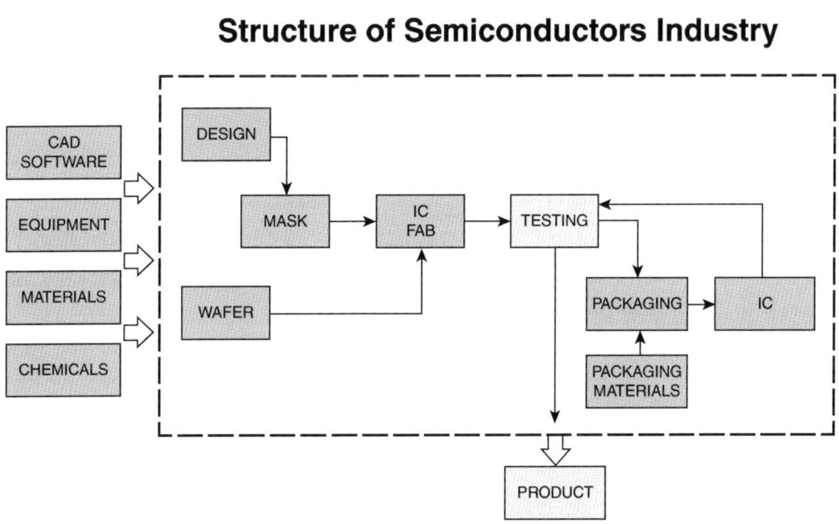

Figure 22. The Structure of the Semiconductor Industry

relationships between the various components of the semiconductor industry. It also highlights the central position of wafer testing.[135]

Linking the front and back ends of the semiconductor manufacturing value chain required thorough knowledge of microelectronic fabrication and possession of the most updated testing instruments. The semiconductor business was intensive in both knowledge (soft power) and equipment (hard power). This was the reason a rather large initial investment was needed.

Another way to highlight the strategic importance of this central position was to look at the front-end of the IC manufacturing process which could be viewed as consisting of three parts.[151]

- Wafer fabrication: pattering the circuitry onto individual dies and repeated many hundreds of times on a silicon wafer. This was a very complicated process that could involve hundreds of steps. It was the crux of "front-end" fabrication and highly capital intensive.
- Wafer probing: examining individual dies on the wafer to check for any defects to ensure that all of the structural parts were in order. Probe testing the circuitry was part of the IC design. As IC chips got more complicated, the testing programs became more stringent. Bad dies were usually marked and the locations were noted on a wafer map corresponding to the wafer.
- Die cutting: cutting the wafer and harvesting the good dies on a plastic or ceramic package through bonding wires or solder bumps.

Finally, these were to be added by testing the packaged IC for all of the designed circuit functions.

In this scheme, testing the packaged IC for every function was a comprehensive and costly process. At this point, dies had gone through the complicated and expensive processes of gold wiring or gold bumping. Thus, some firms held tested dies in a "die bank" and will not package or conduct final testing until the end-use application calls for it.

Third-party Testing Service Providers

Wafer testing is an integral and necessary part of the semiconductor manufacturing process. It is logical to ask why IDMs and IC foundries would leave this important strategic component of its business to third parties, such as Ardentec, instead of doing it in-house to maintain full control of the manufacturing process and the end products. There are several reasons.

First, a paradigm change of the IC business had taken place. In the past, IC transactions were conducted based on the number of wafers, a process referred to as "wafer buy." Each wafer consisted of several hundreds and even thousands of dies, some of which could be rendered defective for various causes during the manufacturing process. The purchaser had to sort out and discard the defective dies, thereby raising the cost of the dies.

In the more recent buying scheme, a purchaser paid only for the number of good dies on a wafer, a process referred to as "good die buy." Therefore, every die on the wafer had to be tested and the defective ones identified and discarded. Updated and sophisticated testing instrumentation required heavy capital investment, generally in the neighborhood of 20% of the total capital outlay. Thus, IDMs or IC foundries had a good reason to outsource the testing service and to focus their own capital investments on process equipment, which was at the heart of their businesses.

Second, third-party services were desirable because they acted as honest brokers between IC manufacturers and IC purchasers. Usually manufacturers and purchasers agreed on the testing program and quality criteria. Testing providers independently and dutifully executed the program and reported the number of known good dies on a wafer. It is impartial and no question needs to be raised and disputed.

Third, the advent of internet technology made this scheme feasible. Test results are immediately available to the parties involved. Any problems could be dealt with instantly. Thus, both manufacturers and purchasers can consider testing service providers as their own virtual laboratories. Parties in different time zones can also communicate seamlessly.

This was the strategic position or the niche advantage that Ardentec identified and developed.

The Vision

Lu and the co-founding team saw the company as a vehicle for lowering the cost of IC manufacturing. Hence, Ardentec's vision was to minimize testing costs and provide the best service to its clients — the IC manufacturing industry. This was a point of fundamental importance.

In 1965, Gordon Moore, a pioneer in microelectronics and an Intel co-founder, postulated that the number of transistors that could be packed onto a typical silicon die would double every year. This claim was found to fairly accurately describe the progress of microelectronic technology over the last 50 years. Thus, it was referred to as Moore's law. A consequence of the law was that the unit cost of IC chip would plummet every year. However, if the cost of testing remained unchanged, testing would become an increasingly larger percentage of the total manufacturing cost. This would make it unsustainable for the industry.

Thus, Ardentec was dedicated to making IC testing a low-cost operation. It hoped to be integrated horizontally with IC manufacturers as a part of the manufacturing process. In other words, they take the core interest of their customer to heart and make it a mission of their own.

Service of Quality

As a testing service provider, there was no substitution for the trust arising from excellent quality, including accuracy, precision, timeliness and confidentiality.

In an interview, Chi Ming Chang, President of Ardentec, commented, "Our business of wafer probe is a niche market, very focused and specialized. We have strived to maintain leadership in wafer probe and system packaging technology, broaden our service to the toughest quality leaders in IC, and play a role of trust and responsibility in the IC value chain." This summed up the corporate culture of Ardentec well. These were the elements of survival and success.

Joining the ITRI in 1990, Chang developed an outstanding expertise in IC manufacturing. He established the computer control technology (PROMIS) of the IC foundry process. He was the manager of the submicron fabrication process of 8-inch wafers. At the VIS, he was responsible for production engineering and, later, operations planning and supply chain management. He was the champion of Ardentec operations. Both Chin and Deng were close comrades. Chin was known for his management skills and Deng for his understanding of semiconductor markets. In 2017, Chin still served as Chief Technology Officer (CTO) of Ardentec. Deng left the company a few years early to lead other major commercial ventures in the Chinese mainland.

The first two years of business operations were difficult for the company. It was a new field for all of the founders. Using the professional reputation and networks they had established from their ITRI and VIS services, they gradually gained repute and acceptance in the field.

However, capital investment was heavy and the capital cost high. Ardentec was in the red in 2000 and 2001 before breaking even in 2002, its third year of operation. Overall, it had experienced a 50% growth in revenue during the first five years. From 2000 to 2015, its CAGR was 22.5%. In 2016, its revenue reached over NT$6 billion. Since 2006, it had expanded business to Singapore, followed by Pyeongtaek (Korea) and Nanjing soon afterward.

Ardentec offered wafer tests of almost all kinds of ICs: DRAM, static random-access memory (SRAM), Flash memory, logic, mixed signal, analog, system on a chip (SOC) and radio frequency integrated circuit (RFIC). Approximately 65% of its customers were IDMs for personal computers, smartphones and other network and communication products. The remaining 35% were customers from IC foundries and fab-less manufacturers. Their customer base extended well beyond Taiwan and into Asia, America and Europe. Many customers would want their relationship with Ardentec to remain confidential until they see fit to make disclosure themselves.

Service Innovation in Knowledge

Given Ardentec's status as a third-party service provider, its relationships with its clients varied according to different needs and circumstances. In addition to performing as an outsourcer, it allowed IC manufacturers to rent testing instruments and provided them its expertise as consultants. Its fees can either be on a per-wafer basis or on die probing time.

As nanoelectronics developed, the circuit line became thinner, the die became smaller and the number of dies per wafer increased. Testing services were more challenging, but prospects were bright. The company recognized that testing service providers needed to be innovative.

One area that Ardentec nurtured as its competitive advantage was its strength in information. This was based on its vision of the horizontal value chain. Lu believed that the detailed data on wafer testing could be of great value to customers. By analyzing the test data, they could optimize their production process, maximize the yields and improve the product design.

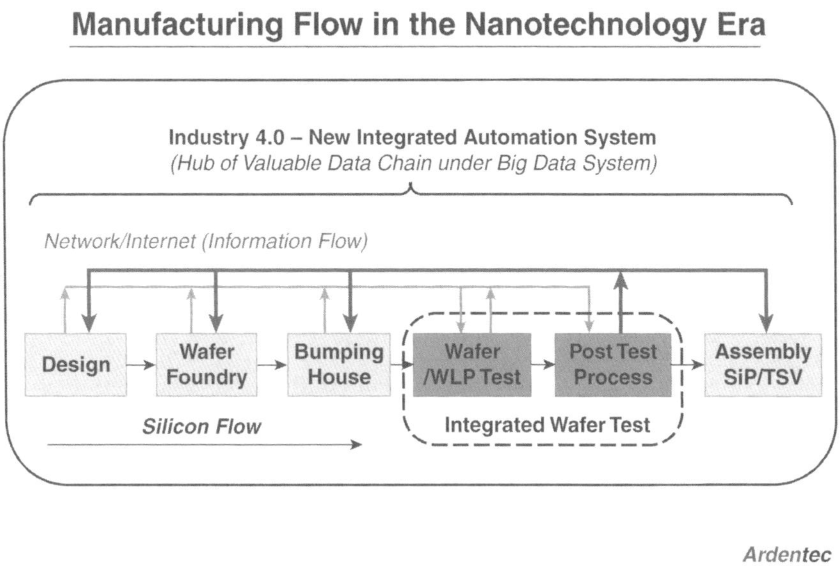

Figure 23. Testing as the Hub of Manufacturing Information

By linking to the internet, information could be transmitted to customers in real time. Thus, providing an information flow would create insightful value for the customer. This is shown schematically in Figure 23.

This value was anticipated to be more significant for high-level electronic packaging, such as SOC, system in package (SIP) and multi-chip packaging (MCP).

According to a recent ITRI analysis, the top six added values of industrial products from Taiwan were ranked as follows:

IC Foundry	71.4%
IC Testing	55.9%
Display Panels	26.7%
IC Design	22.2%
LED	19.1%
Photo Voltaic	8.9%

This highlights the importance of the IC testing industry as a whole.

Advanced IC testing is a very specialized business. There are no raw materials, no manufacturing, no inventory and no pollution. It is a business of knowledge and knowledge application. Customer trust is earned through continuous high-quality service. In my discussions with the co-founders, I was particularly pleased that the ITRI's culture of innovation, teamwork, respect, diligence, dedication and excellence was affirmed in the IC testing business. I was gratified that the co-founders of this young company have reconfirmed these values well through practice on the job. This was true to Confucius' teaching of "loyalty to job, trust to friends and practicing what one learns."

Epilogue

The following press release was made by Texas Instruments Incorporated in Dallas on April 4, 2017, through *PR Newswire*:[154]

"Sixteen companies have been recognized by Texas Instruments Incorporated (TI) (NASDAQ: TXN) as winners of its annual Supplier Excellence Award (SEA). As the company's highest level of supplier recognition, the award highlights suppliers who have excelled in conducting business ethically while providing outstanding products, service and support that meet or exceed expectations for quality, environmental and social responsibility, technology, responsiveness, assurance of supply and cost. The sixteen winners are an elite group selected from more than 12,000 suppliers.

'At TI, we know our customers depend on us to help them innovate and get to market first, and we have the same expectations of our suppliers,' said Rob Simpson, vice president of TI Worldwide Procurement and Logistics. 'The Supplier Excellence Award winners have demonstrated exceptional commitment and capability to deliver the products we need and the performance we expect.'

The sixteen SEA recipients, along with the products or services they provide to TI, are:

Ardentec — Test services

(and fifteen other semiconductors companies worldwide)"

…

Coming from a customer, this is a fitting appraisal of Ardentec Corporation and its entrepreneurial culture.

Chapter 23

DELTA ELECTRONICS: PARTS AND COMPONENTS

Delta Electronics was established by Bruce Cheng in 1971 in Taipei with under US$10,000 and a staff of 15. It produced television coils and intermediate frequency transformers (IFT) for local and global television makers. It was established as a high-quality manufacturer. In 1978 it won the Outstanding Quality Award from the Radio Corporation of America (RCA) for supplying 20 million components over two years with no defects. It continued to win awards from the RCA and Zenith almost yearly.

In the mid-1980s, I began working at the ITRI in Hsinchu, first at the new MRL and then at its headquarters. I met with Cheng a number of times and was quite impressed with the vision he had for his company. We both felt it was mutually beneficial to foster cooperation between the two organizations.

However, it was not until 2016 that I had the opportunity to study more deeply his entrepreneurial odyssey.[136] *I was curious about his decision to give up a well-paying job with a famous foreign firm, the envy of young engineers at the time, to embark on the risky and tumultuous road of entrepreneurship. I also wanted to assess the factors contributing to the success of his entrepreneurial career.*

A Snapshot

In 1984, Delta Electronics started making switching mode power supplies and electromagnetic interference (EMI) filters for personal computers manufacturers. Its client list included Acer, Digital, Philips, Rockwell, Wang and Xerox, among others. As a result of its performance quality, the business quickly expanded. In 1988, its turnover exceeded US$100 million.

From 1990 to 1996, its product lines expanded to color monitors, communication power systems and AC motor drives. It won vendor awards from Acer, IBM, Epson, Hewlett Packard, NEC, Philips, Intel, GE, Wang and Xerox. Its plant in Taiwan received ISO 9000 certification. The company then established factories in Mexico, Europe and mainland China. In 1996, its turnover exceeded US$1 billion.

From 1997 to 2000, it moved on to green batteries and energy systems. The plants in Taiwan and China received ISO 14001 certification. The company's sales network became global. In 2000, its turnover reached US$2.5 billion.

In the new millennium, the company strived for higher technology and innovative content in its product lines. From 2001 to 2005, it won awards for science and technology from Intel, Sony, Microsoft, Cisco and Siemens. In 2005, its turnover exceeded US$3.5 billion.

Since 2006, it had further moved into green energy and environmental areas. It was a key technology supplier of the 100 MW solar power plant in the western U.S. desert. The plant covered 6.48 million m^2 with over 10,000 $10.5\ m \times 10.5\ m$ solar panels. The electricity it generated was used to power a city of 75,000 residents.

Delta Electronics had been cited by many global journals and institutions, such as Business Week Info Tech 100, Global Top 100 Low Carbon Pioneers and Forbes Asia's Fabulous 50.

In 2014, its consolidated worldwide sales were US$7.52 billion. Over the 40-year period since its establishment, its CAGR was 32.6%. It was a global company also known for its forward-looking corporate culture, which valued innovation, quality, agility, teamwork and customer satisfaction. In 2014, Cheng was elected Honorary Chairman of Delta Electronics.

Childhood in Wars and Suffering

In July 1937, when Japanese troops attacked the Marco Polo Bridge 50 km to the southeast of Beijing, Cheng, born Chong-Hua (鄭崇華) meaning "the exalted China", was just one year old. The Japanese attack formally started the Sino-Japanese War. It brought many abrupt changes to life of the Fujian Province town where his family lived, approximately 2,500 km south of Beijing.

The town, although small and quiet, had a depot that could presumably be used for aerial transportation. Therefore, it became a target of Japanese air raids. Japanese warplanes first dropped bombs with explosives, then later with rocks. Unknown at the time, the rocks were smeared with bubonic plague bacteria to cause an epidemic in the general area. These random bombardments created so much havoc that Cheng's parents decided to move the family to the countryside to join his maternal grandparents.

There, farming life was relatively tranquil. The guidance of a learned grandfather and the mentorship of both parents have provided a warm habitat for the growing boy. Bruce's father was a devoted practitioner of traditional Chinese medicine and his mother a caring schoolteacher. Thus, "Like a blessing in disguise," grinned Bruce Cheng as he reminisced about his childhood, he was able to grow up in an environment ingrained with Confucian values: diligence, trust, thirst for knowledge, respect for ancestors, love of nature and animals, striving for excellence and social responsibility. These values were important factors that defined Cheng's entrepreneurial career.

Approximately 10 years later, China was immersed in war again. This time the fighting was internal, between the rising Chinese Communist Party (CCP) and the ruling nationalist government under Kuomingtong (KMT). A young uncle on Cheng's mother's side brought him along as he traveled to Taiwan in 1948. The refugee-like lifestyle had trained Cheng to be perceptive, perseverant, thrifty, independent and self-motivating. It also nurtured in him the ability to build support network with friends.

After high school, Cheng entered the National Cheng-kung University in Tainan, first majoring in mining and metallurgy before transferring to electrical engineering. He greatly enjoyed his studies and life at the university. He was especially inspired by his chemistry professor, Lai Zai-de, who taught him that man studied to acquire knowledge, not to pass exams. This has helped him to focus on his priorities in life.

Learning Respect for Job and People

Cheng graduated from college in 1959 and went into military service with the Air Force (as part of the Reserve Officers' Training Corps). After completion of the service, he married his long-time girlfriend and got a job at the instrument shop of Air Asia's hangar facility in Tainan.

His responsibility was to repair and maintain rather sophisticated airborne instruments. This job provided him with many technical training opportunities and exposed him to high-performance instrumentation. It made him realize the vital importance of product quality. He benefited from his American supervisor, who mentored him on not only technical skills, but also management ability. Gradually, he became able to design and make improvements to normal equipment with ingenuous functions.

Although his job progressed nicely, tensions were growing between the foreign managers and local employees. The management showed little respect for the locals, hiring and firing at its own discretion. Job demands had also become increasingly unreasonable. Cheng was highly regarded and was often asked to do jobs for which he was not properly trained for. This made him uneasy because of the risks involved with product safety and public liability, the ramifications of which were formidable.

Cheng contemplated changing jobs, despite having good career prospects and a generous salary. His five years at Air Asia gave him a first-hand understanding of the importance of creating reliable and high-quality products. It also gave him the confidence to handle difficult assignments and the art of working with people.

A Taste of Foreign Enterprise

Starting in the 1960s the nationalist government in Taiwan, under the leadership of Chiang Ching-kuo, Yuen-suan Sun, Kwoh-ting Li and others, embarked on aggressive measures to boost the economy. Policies and programs were instituted to encourage foreign investments, especially those promised to help boost the nation's technology and management capabilities. TRW was one of the companies that came to Taiwan. It focused on manufacturing electronic components, such as transformers for television sets and variable capacitors for car radios.

Cheng was recruited to become TRW's Production Manager and was sent to the U.S. for training. This broadened his knowledge of product design, production control and industrial engineering. His stay in TRW's Waseca, Illinois facility was a valuable experience. He started to appreciate the warmth and kindness of rural Americans. The thought was especially touching to him, as the factory people there knew very well that after the training, the trainees were going to set up operations to compete with them.

After Waseca, Cheng spent time at TRW's Chicago plant, which gave him more insights into the different factory cultures within a big company.

After returning to Taiwan, Cheng was quickly put in charge of running the production line. This was a very different job from his previous one at Air Asia, which was oriented toward problem solving and R&D. Here, he kept an update of various parts and components from many suppliers; whose quality and quantity varied greatly. With rapid market expansion of the television industry, demand for TRW products increased so rapidly that U.S. suppliers could no longer keep up with the parts needed in the Taiwan factories. Thus, they had to source from local suppliers, whose product quality varied.

Over time, Cheng established a network of qualified suppliers which enabled TRW to reduce costs. As the business progressed, Cheng became a member of management to work on the company's five-year plan. This allowed him to participate in goal setting and management and eventually take charge of quality control for all four TRW plants in Taiwan.

Cheng learned that to do the right job with high quality, he had to get things right the very first time. He was gradually exposed to various quality techniques, such as the Japanese methodology of zero defect management.

At that time, standard TRW quality control practice was a 10% sampling of products. Cheng thought this inadequate and took the initiative to move up the quality ladder. By enforcing the Military Standard Sampling Plan, he was able to tighten the production line. However, more production rejects at the end of the production line did not make him more friends in the factory. Fortunately, top management supported his actions. This confirmed that commitment from top management and adequate training and communication with peers are essential in the corporate quality performance efforts.

Motivation for Entrepreneurship

TRW management practices at the plant became increasingly troubling to him. Employees were often fired when business was slow and hired when business returned. Everything was done to cut costs and increase profit. This practice extended to all levels of local management, as higher salary jobs were easier targets for cost reduction. Cheng felt that this was grossly unfair to the employees, disrupted the business and was much more costly to the company in the long run. He argued his views to no avail. Furthermore, there was little or no intention or willingness for the company to establish its own technology base in Taiwan. Neither was it prepared to help enrich the community and nurture local workers whose efforts made the company prosperous. This gave Cheng many sobering thoughts for the future.

After working for TRW for approximately five years, Cheng made up his mind that he wanted to build a company that will treat employees with respect, develop unique technology and offer opportunities for local talents to fulfill their potential. He believed that on this basis, he can make excellent products that bring pride to Taiwan. This was what Delta Electronics set out to accomplish.

The Opportunity

In the early 1970s, there were several manufacturers in Taiwan that produced and exported television sets to the world markets. The market share was small, approximately US$1 billion, but substantial for Taiwan's rising economy. Major producers, such as Tatung and Sampo, started their businesses by selling imported television sets to domestic markets. Not having their own technology, they bought kits of components from Japan, where they sent their engineers for training, and returned to Taiwan to assemble the products.

As business grew, these companies wanted to establish their own design capabilities; but their Japanese technology source was reluctant to support. They then turned to other companies, such as TRW Taiwan, to source parts and components. Initially, TRW had some interest in this, as its own business had experienced low cycles. However, such relationships did not materialize because the quantities needed were too small to be economical for TRW.

Cheng was the TRW engineer who worked with Tatung during this process. Although the company's formal business relationship with Tatung fell through, Cheng was sympathetic and helpful and was often consulted on technical problems. Thus, Tatung was very pleased that Cheng was establishing his own company. They trusted Cheng for his personality, knowledge and willingness to cooperate for mutual benefit. This set Delta Electronics off to a good start.

Indeed, the company was able to give Tatung good advice on many new designs. It provided not only good quality components, but also very competitive prices and prompt service. Very shortly, Delta Electronics was also qualified by the RCA and Zenith and was able to provide parts to global television manufacturing giants for not only their Taiwan factories, but also their U.S. headquarters.

Challenges of Entrepreneurship

Cheng soon realized that knowing technology alone was not enough to run a company. Challenges arose from so many aspects of the business that a technical person never would have expected. For example, one of these was the management of cash flow.

Tatung was a very good customer and its business volume was profitable. However, it maintained a long payment cycle. Accounts receivable often varied over 30, 45 and 90 days. And, whenever its chairman traveled abroad, the accounting department did not allow checks to be issued, as they lacked his signature. Fortunately, Taiwan's government instituted loan guarantees to qualified SMEs, which provided some relief.

The cash flow problem was not entirely solved until transactions with large U.S. companies started. Generally speaking, the financial terms with global companies were clearly outlined and, thus, more manageable. With letters of credit from these companies, local banks promptly provided short-term working capital. Cheng said, "This is one of the advantages of being a player in the global market: you play by the rules of the market." Reminiscing about those trying times, he further commented, "Although I have gone without pay a few times, I have never missed a payment to my employees."

Quality as Competitive Advantage

From the beginning, Delta Electronics was guided by its quality principle of doing things right the first time. As a newcomer, this was how it was able to forge business relationships with the RCA and Zenith, becoming the supplier of not only their Taiwan subsidiaries, but also their global manufacturing sites.

Enforcing good quality assurance became a standard practice of the entire management team. It was reinforced by adopting advanced technology, such as automation and military specification methodology which makes each product effective and traceable.

In anticipation of changing industrial needs, Delta Electronics was the first manufacturer of electronic components in Taiwan to use surface mount technology. With this, it was able to participate in the early development of printed circuit boards for disc players (Qume) and desktop computers (Acer).

Up the Value Chain

The ability of Delta Electronics to leap forward to the leading edge of the industry became its competitive advantage. Having developed a solid position in electronic components technology for the television industry, Cheng started to map his strategy for the next decade.

In the early 1980s, he became interested in the growing trends of the computer industry. He knew that his core strength was in electronic parts and components. He felt that Delta Electronics could enter the computer arena from the angle of parts and components, which all computer manufacturers must need. The company eventually became a valued partner for many brand name players. Capturing new opportunities from a position of strength was indeed a very sound strategy.

Thus, Delta Electronics started producing switching mode power supplies and EMI filters. The filter minimized or eliminated the electromagnetic interference of components which greatly improved computer stability and performance. This was a must for all computer makers. A switching mode power supply was a switching regulator that converted electric power efficiently from an AC or DC source to DC loads in personal computers. It minimized power dissipation and generated the least amount of heat. These features were extremely important for any PC, as they were to be made lighter, shorter, smaller and thinner, and the problems will become more acute. Delta Electronics quickly became a supplier for Acer, Rockwell, Digital, Wang, Xerox, Hewlett Packard and IBM. In 1988, its turnover exceeded US$100 million, which was a landmark for a start-up specializing in incremental and continual innovation.

Challenges of Globalization

Recognizing its need to serve many customers in different business areas worldwide, Delta Electronics started its globalization efforts. It first established sales offices in the U.S. in 1980, followed by offices in Europe, South America, Asia and the Chinese mainland. It started from sales, services and manufacturing before moving on to technology development. Although the choice of globalization was certain, the challenges of globalization were many. Issues concerned with were human resources, financing, logistics, law and integration.

Many companies had taken similar approaches and encountered similar problems. Delta Electronics learned to handle problems very carefully. Cheng recognized that the most important factor to successful globalization was addressing cultural differences.

Some salient cultural differences could be noted, based on his own experience. For example, workers in Mexico were friendly and enthusiastic, but easily jumped to rival companies for slightly higher salaries. Malaysian workers were smart and attentive, but many had biases against foreigners, especially the Chinese. Thai workers were very happy and easygoing, but were generally laidback and unwilling to work a few extra hours, even with pay. Workers in Germany were disciplined and innovative, but often seemed rigid and aloft. French workers were creative and resourceful, but many were self-centered and worked at very slow pace. In Japan, people were courteous and diligent, but also reserved and suspicious.

These observations were not meant to be generalizations. They merely indicate that people are different in temperament and lifestyle, as Cheng observed. Therefore, company management should respect such differences and take different actions to lead and motivate the employees. Cheng commented, "If you want to sell a product, open an office or build a plant, be prepared to execute your plan differently in different locations."

As of 2015, Delta Electronics had established approximately 100 operational sales offices, 16 manufacturing plants and 35 R&D centers worldwide, excluding China. In the Chinese mainland, it established approximately 50 offices, 20 factories and 20 R&D centers. Globalization has become an integral part of its corporate strategy.

Learning from his own experience working with a foreign company in the early days of his career in Taiwan, Cheng established guiding principles when Delta Electronics moved abroad. The company did not pursue globalization merely to take advantage of cheap labor and low costs. It should maintain an attitude of treating local people as overseas partners and focus on stimulating local aspiration for the future. It allowed those local companies capable of designing their own products to do so with full support. It never sought to take layoffs lightly, let alone pull out investment all together. Finally, it set out to foster a strong sense of loyalty among its local employees. These guidelines reflected the company's corporate culture.

Focusing on Energy and Environment

Being in the business of electronic parts and components, Cheng developed a sharp sense of energy efficiency. In the 1970s and 1980s, as Taiwan became increasingly industrialized, electricity usage grew at an annual rate of 7%, straining the country's power supply system. There was strong pressure for the government to build more power plants. Cheng felt it would be a lot easier and more cost effective for manufacturers to make energy-efficient products and for households to use them. Without a conscientious effort in energy conservation, there would be energy crises, if not now, but most certainly in future generations.

These issues were behind his efforts to manufacture switching mode power supplies. He knew very well that the demand for energy and environmental sustainability was going to be a future trend. By switching mode power supplies, he could accomplish several objectives. Indeed, Delta Electronics' SPS products had been widely welcomed by manufacturers. In the mid-1990s, its product lines occupied approximately 50% of the world's market share.

With this vision, the company continued to produce energy-efficient and environmentally friendly products. In 2004, it moved into the solar cell area, followed by LED array projects for green energy and carbon reduction purposes. The company was eventually organized according to three business categories: Power Electronics, Energy Management and Smart Green Life. In 2007, Delta Electronics won Forbes Asia's Fabulous 50 Award. In 2008, it was listed as a Global Top 100 Low Carbon Pioneer.

Leveraging on Technology Institutions

Knowing that it would be difficult to achieve its goals and business objectives through its own R&D efforts, Delta Electronics collaborated with leaders in technology. In various phases of its development, it worked with many universities and institutions in Taiwan and abroad.

A significant effort involved its collaboration on switching mode power supplies with the Virginia Polytechnic Institute under the direction of Dr. Fred Lee. In 1989, Delta Electronics established a power electronics laboratory at the institute's Electronics Center. It later moved to Research Triangle Park in Raleigh, North Carolina, where it established the Delta Network Research and Development Center. Its efforts in power electronics were continued later in China with Tsinghua University, Nanjing University, Zhejiang University, Xi'an Jiao-tong University and others. All of these efforts helped make Delta Electronics a leader in power supplies and power management. Another project was conducted on artificial intelligence and speech recognition in conjunction with Dr. Victor Zue of the MIT. Cheng personally spearheaded many of these projects.

In Taiwan, Delta Electronics conducted many projects in collaboration with the ITRI, such as thin film and solar cell projects. Its engineers worked closely in the ITRI's materials, chemicals, energy and electrical engineering laboratories. It also collaborated with universities on human resources research and training. Furthermore, it made substantial donations and supported academic exchange programs at the National Cheng-kung University and other institutions. Overall, enthusiasm for technology and a strong sense of social responsibility made Cheng and Delta Electronics active contributors to the transformation to a knowledge-based economy in Taiwan.

Chapter 24

McDONALD'S HONG KONG

On a humid autumn night in Clear Water Bay, I toss and turn in my bed, trying to get some sleep. My mind keeps wandering to a recent discussion with the President of the university. Oddly enough, as a career university professor, he has attempted several times to persuade me not to teach any courses. Why? He says, "Your position at the HKUST as Vice President entails full-time engagement in administrative duties. Preparing for courses and conducting discussions would detract your attention and take time out of that required by the position."

I protest strongly that this university has attracted me not because of its fabulous ocean view or liberal compensation package, although I must admit they help. Rather, I am attracted to the opportunity to engage in intellectual interactions with the faculty and students in an academic environment. Otherwise, sitting and debating bureaucratic matters in meetings the whole time is hardly stimulating. It would quickly reduce me to the likes of a stale walnut, defeating the purpose of having me on the staff.

In the end, he concedes on the condition that I teach one course in the evening, but that I continue to work full-time during the day on the duties outlined by my contract.

I feel a special burden to prove that my interest in teaching would benefit the students and the university. But, how?

Just as soon as I feel drowsy, I am awakened by the telephone. It is the on-duty security guard, who says, "Professor Lin, we want to report that a Number 8 typhoon signal has been hoisted." "Uh? Thank you," I respond. "Well then, the university will be closed tomorrow. There will be no class or official business." Somewhat consoled, I quickly catch up on my sleep.

"Ring, ring, ring!" I am soon jolted by loud noise and grab the phone. It is barely a few minutes past 8 o'clock. A strange voice is coming from the other end of the line, "Otto, this is Daniel Ng. We have an appointment today."

Hearing this name fully awakens me. I say, "Yes, indeed, Daniel. But, don't worry. As the typhoon is still with us, we can arrange for another time to meet." He says, "Actually, I am already on campus."

I spring up from the bed and ask, "Why, how so?" He calmly explains, "It was a Number 8 typhoon signal, so the road was almost empty. I took the wheel and found myself at your campus in no time."

Entrepreneurship by Example

Under a rather unusual conditional agreement, I started teaching "Management of Technology" in the evenings at the School of Engineering in 1999. The course was aimed to teach engineering students the basics of management. I thought it was logical for a young engineer advancing in an organization to eventually become a manager. Therefore they need to prepare for such a career challenge even though management is not a choice of their major study. As most of Hong Kong's manufacturing activities had relocated north to the PRD or further into the Chinese hinterland, a practicing engineer in Hong Kong is going to enter the management field, sooner than expected. He will face many cross-boundary, multi-cultural, multi-client and techno marketing issues in the globalization era. The course focuses on the various management functions in practice, not on the theories of management science.

In the course syllabus was the subject of entrepreneurship. I viewed the entrepreneur as the key person in championing innovations to create value and wealth. Although the course covered many aspects of entrepreneurship, I thought it was desirable to invite model entrepreneurs to share their experiences with the students. I hoped the students could learn from them as people rather than just names in a case study book.

Of the many possible speakers, I wanted to invite Daniel Y. C. Ng, the founding CEO of McDonald's Hong Kong. I had always advocated that innovation entailed not only novel product and process technologies, but also business models and business plans. McDonald's restaurants fit my description.

The case of McDonald's in Hong Kong caught my attention particularly. Its products were everyday food items — nothing fancy. I was interested in learning how, almost 25 years ago then, the founding CEO decided to sell hamburgers in Hong Kong. What efforts did he make to persuade the taste-conscious Hong Kong locals to like the grainy and bland-tasting beef patty? This seemed especially challenging. Around every street corner were shops that offered sumptuous roasted pork, broiled duck, steamed fish and various kinds of dim sum, all at very reasonable prices. The odds against this venture seemed formidable.

The Moves of an Entrepreneur

I had no prior knowledge of Ng, nor had I ever met or communicated with him. Frankly, I doubted he knew me either. Tracking him through the company route was a real challenge. Nobody knew exactly where he was. He had already retired from McDonald's, this was certain. But very little else beyond that. He was known to travel extensively. It was rumored that he was flying helicopters around Turban and Urumqi in Northwest China on some new venture! This turned out to be true. He was then engaged in promoting ethnic music and culture and became an important member of the Silk Road Project founded by the renowned cellist Yo-Yo Ma.

Finally, a friend gave me an e-mail address, but could not guarantee that it was current. I wrote a courteous letter, introducing the university and the course and inviting him to be a guest lecturer. As expected, I did not receive a response. For almost two months, I repeatedly sent the e-mail to him, each time with a tone of added urgency. Finally, a response came. He wanted to visit the university, learn about me and discuss what he can could really do for my class. I was overjoyed. We agreed on a date after his return to Hong Kong, which fell approximately one month later.

The night before this date, the Number 8 typhoon hit Hong Kong, and all public offices and schools were closed. However, Ng remembered our appointment. He drove to the east end of the Kowloon Peninsula by himself in adverse weather conditions, an inconvenience to say the least.

A Vision of Corporate Value

As I hurry to the campus atrium, the wind is whistling and raindrops are twirling all over the hall. I see a man with a raincoat on his arm examining the surroundings with curiosity. I dash forward, extend my hand and ask, "Daniel?" He smiles and catches my hand. "Good morning, Otto."

I ask, "Why didn't you sit inside the coffee shop on the other end of the hall?" "Well," he says, "I did try as you suggested, but the shop is closed."

"Um, in that case, let's go to my office. But you will have to settle with the coffee I make." As we walk to the elevator lobby, I say, "On second thought, perhaps we can try the McDonald's on Lower Level 5. One has recently opened there. But I suspect it may also be closed."

He looks surprised, "Oh, you have a McDonald's on campus? If you have a McDonald's here, it must be open."

"How do you know that? It is a Number 8 typhoon signal after all."

"I know because I set the rules. All McDonald's restaurants must stay open in bad weather, even Number 8 typhoons. Precisely at a time like this, there are people who need a place to sit and eat. We are here to serve the people. How can we be absent when they really need us? All store managers must make sure to get their employees to the store. They would order a taxi or take the wheel themselves if they needed to."

"Well, it is a very good thought. But this is a new store. Let's go and see." We take the lift down to Lower Level 5. I look to the right and there it was, the big bright shining yellow "M" sign. Many customers were already there, contentedly sipping their coffees.

As we walk into the store, a man behind the counter appears to be thunderstruck. He hurries over, vigorously shakes Ng's hand and says, "Good morning, Mr. Ng. What a pleasant surprise to see you!"

I am quite moved by the scene, but cannot help thinking that the man almost had a heart attack. He must think the retired Chairman had come to conduct an unannounced inspection on account of the weather conditions.

I admit to Ng that as we have never met before, this is a positive identification of him, which is mighty good for me. He laughs heartedly. As we settle at a table, we both seem to savor a moment of comfort while the wind blusters outside the window. Then, with the aid of McDonald's signature coffee, we move on to the subject of his founding McDonald's Hong Kong, and his interest in guest lecturing my class on entrepreneurship.

Strategic Focus: People

I start by asking the question that has been on my mind for a long time: "How did you go about persuading Hong Kong people to eat hamburgers some 25 years ago? Noting that at that time, not too many local people, even in Hong Kong, the supposed gateway city of East and West, had traveled to the U.S. or been exposed to American-style cuisine."

"Actually, selling the beef patty as a product is only a small part of the business. We asked ourselves, how can we attract and serve the people who come to our restaurant and make them return with their families time and time again?"

"Uh, people are the strategic focus; the real 'beef' of the business!" I exclaim.

"Yes. What we are really selling is our service with high standards: value, quality, speed, cleanliness and courtesy. When the customer feels comfortable at our place, whether he wants a hamburger, chicken, coffee or Coca Cola is his choice. He may try one this time and another the next time."

He continues, "Our customer knows that when he comes to our store, whatever he buys is worth the money. This means value. Our meal has consistent and reliable quality. The service is quick. We prepare the food while customers wait. Our restaurants are clean and our crew are courteous. The first words our customers hear are 'welcome' or 'good morning' and the last words are 'thank you' or 'come again.'"

As he describes the scenario, I immediately picture going to a local, normal, traditional restaurant of the same price range. The tables are generally greasy, the floors gluey, the waiters hasty and brusque and the quality of the food variable. Overall, you never know when you are going to be served. When an order finally arrives, you will have to hurry to finish it. If you sit and relax for a bit too long, your waiter looks at you with questioning eyes. What a contrast!

Business Growth

I started to realize the importance of Ng's business strategy of focusing on servicing the people. However, much of this hinged on the workers in the restaurant. How could you ensure the employees were going to do what you wanted them to? "Training, obviously."

Training was a very important part of McDonald's. The company did a lot of local training for its employees. And, in fact, the U.S. headquarters ran a university for training people from different job levels. There were approximately five of such institutions worldwide, dubbed "Hamburger University."

Equally important, if not more so, was that all of the employees felt they were part of the company and respected by everyone there. According to Ng, "We call our employees the 'crew.' Just as in a sailing ship, we are all in the same boat." In the majority of restaurants in Hong Kong, employers and managers are usually addressed as "Bossie" (波士 or 佬細) and employees are simply "Hey you" (伙記). At McDonald's Hong Kong, everybody was addressed as the crew. Therefore, the employees felt respected and that they were part of the company. As such, they gladly accepted training and acted like a member of the ship.

Ng opened his first restaurant in Causeway Bay in 1975, on a very busy downtown street. Thousands of people walked past the restaurant every hour. If 3% stopped in, he will have a full house for the entire day. When a customer came in, he will surely find the McDonald's experience very different from other local restaurants. Through word of mouth, the new restaurant became a tremendous success. By the time Ng retired as Managing Director, there were approximately 200 branches in Hong Kong employing approximately 8,000 crewmembers.

The Risks of Entrepreneurship

Before discussing the university in Hong Kong and my class, I ask another question: "Had you considered the odds against this venture back in 1975?"

"Certainly, indeed. A lot of people thought I was crazy. The fact was that as more people voiced doubts about it, the more it was desirable to me. This is how I saw it. The fewer the people who wanted to get in, the freer I would be to take action and the greater the opportunity."

Ng was trained as an engineer. His career path as an entrepreneur was rather legendary. He was born and grew up in Hong Kong and later studied in the U.K. He received a doctoral degree in chemical engineering from Imperial College in London. After graduating in 1965, he joined the Institute of Gas Technology in Chicago to conduct research on energy and the environment. He returned to Hong Kong seven years later to get into the venture capital business. In 1975, he started the McDonald's Hong Kong venture.

Ng was a person of immense creativity and independent thinking. One of his interests was music. He had long believed that the Silk Road was the role model of multicultural exchange. This prompted him to join the Silk Road Project to pursue his long-term interest after retiring from McDonald's Hong Kong.

On lecture day, the students responded most enthusiastically to Ng's presentation. As an introduction, I told the story of the wake-up call on the morning of the Number 8 typhoon. I thought it showed the basic personal characteristics of an entrepreneur: diligence, confidence, courage, ability to keep promises and take risks.

For the People

As we discussed Ng's lessons, many subjects came to mind. The first was placing a strategic focus on people. Although McDonald's provided a real and visible physical product, such as a hamburger or a coffee, it was the value created and the service provided during the process that set the restaurant apart from others. When the focus was on people, efforts were made to ensure that customers felt good at the restaurant. The customer's interest came first. He was not to be rushed and could stay as long as he needed. He was not to be pushed, and any order was his own to make.

In a more recent discussion, Ng mentioned his admiration of the Apple Store for practicing the "people first" principle. At the Apple Store, attendants were all well-prepared problem solvers and provided very friendly service. They did not promote or pressure the sale of any product, large or small. It was all up to the customers.

A consequence of taking the interest of customers to heart was that local restaurants had some flexibility in arranging the decor and formulating new products tailored to locals' interests. Cleanliness was also a distinguishing feature. Seats, tables, public areas and restrooms were all nicely kept. This was one of the features in which McDonald's excelled in comparison with other restaurants in Hong Kong. Thus, a visit to a McDonald's was always an enjoyable experience. It was also a top choice for family outings, especially with members under 10 years and over 60 years of age.

Technology Conscious

Up-to-date technologies were used to support the value chain. The foods McDonald's offered were the everyday items that Americans had eaten for hundreds of years. However, the food quality was upheld by modern materials processing, quality assurance and global supply chain management. For example, the freshness of steak and crispness of fries were closely monitored. If anyone was unhappy with any of the products they had received, they were replaced with no hesitation. Returned items were immediately discarded, not kept or sold to someone else.

All of the franchises operated under the same business model of onsite manufacturing and service under a quality management system. Understandably, McDonald's was not an organization that developed technology, but one that conscientiously followed the advent of technology and applied it to its regular business operations.

A Role Model

McDonald's set examples for innovators. Many fast food chains established operations in the U.S., as McDonald's became popular after its inception in 1955. The list included Kentucky Fried Chicken (and its later version, KFC), Burger King, Wendy's and Arby's. All of these became standard in the daily lives of average Americans. After McDonald's Hong Kong, the trend was widely reproduced in Taiwan and the Chinese mainland. This was perhaps a development that even Ng had not expected.

Fast food is a culture of our time. The total number of jobs created, sales revenue earned and tax dollars paid to various levels of government are considerable. Some scholars suggested that the influence of the McDonald's Corporation on the food industry could be likened to Ford Motor Company on the automotive industry. Both significantly affected national and global economies. This was truly innovation at work.

In 1986, *The Economist*, invented the Big Mac Index as a lighthearted guide for the value of different currency. Thus a "Big Mac" priced at $5 in Country A and $3 in Country B, exchange rate adjusted, will indicate Country B's currency is about 40% undervalued, relative to A. This subject has aroused a great deal of academic interest. An index adjusted for GDP per capita can show more insight on the fair value of currency.[169] Regardless, it says that McDonald's has become a global phenomenon.

The fast food business model also had unexpected effects on Chinese cuisine, which was generally known for its complex and delicate processes. The success of McDonald's prompted many entrepreneurs in China to simplify and systematize the food preparation process, leading to Chinese-style fast food. In many larger cities of China, Chinese fast food restaurants gradually sprouted. I personally visited restaurants, such as True Kung-fu (真功夫) and General Chopstick (筷將軍), and was quite impressed with the creativity of some innovators. I had not yet tried Beef Noodle Soup of California (加州牛肉麵), which was reportedly an innovation from Taipei that had gained popularity in Beijing and Shanghai. I marveled at the seemingly unlimited innovations in the food and restaurant businesses in the globalization era.

Social Networking

McDonald's played an important role in social networking. Its role took many forms. In some Eastern European countries, going to McDonald's was a status symbol. In the early 1990s in Moscow, I saw people waiting in very long lines outside the gigantic McDonald's restaurant in Moscow. It appeared to be as big as the central railway station terminal. In cities such as Hong Kong, Taipei and Shanghai, McDonald's restaurants were teenagers' favorite places to do homework and meet friends. Many restaurants built playgrounds to attract parents and their young children. McDonald's also became a favorite of retirees because of its relaxing atmosphere and its senior's coffee offer!

All of its development could be understood in terms of its strategic focus on people. This was the underlining principle of entrepreneurship that Ng upheld. The class was an extraordinary success. I truly felt that all of the time and effort I had spent on it were well rewarded.

Epilogue

On January 10, 2017, the McDonald's corporate headquarters announced it was selling 80% of its fast food business in Hong Kong and mainland China for US$2.08 billion to the largest Chinese international trust company, CITIC Capital, and a U.S. private equity firm, the Carlyle Group.[137] After the cash and shares transaction, CITIC Capital was going to own a 51% stake and the Carlyle Group a 28% stake. McDonald's was going to keep a 20% interest in the company. The new partnership was going to be the master franchisee responsible for McDonald's business in China and

Hong Kong. Over 1,500 restaurants were going to be opened on the mainland, targeting the third and fourth-tier cities in five years. At the time, McDonald's owned 2,400 outlets in mainland China and 240 in Hong Kong.

Some business observers believed the deal was based on the strong cash flow and performance of the McDonald's restaurants, despite an increasingly competitive market in the region.

Perhaps Daniel Ng, as the founding entrepreneur, found comfort in his role in starting a great industry and lifestyle in the Greater China Region.

Chapter 25

ALIBABA GROUP

On January 10, 2017, Ma Yun (also known as Jack Ma) of the Alibaba Group (China) and Masayoshi Son of Soft Bank (Japan) met with U.S. President-elect Donald Trump in New York and offered assistance in strengthening American business.[138] *It was reported that "Jack Ma offered the company's e-commerce platform to help U.S. businesses sell products to Asia, with the potential of creating up to 1 million American jobs, giving Donald Trump bragging rights for fulfilling his campaign vow even before he formally [took] office."*

The report ended with, "Trump said he and Ma had a great meeting," adding that "the Alibaba founder is a great, great entrepreneur, one of the best in the world."

What a difference 18 years made. Alibaba.com was established in 1999. It offered a business to business (B2B) platform to help Chinese businesses sell products to the U.S. What was Ma Yun going to name the future platform, "Ali-trumpi"?

The Debut

As a teenager, Ma Yun was fond of the English language. He liked to bike along the tree-lined trail surrounding the dreamy West Lake. With a small transistor radio in his pocket, he whistled to music, followed the words of broadcasts and was happy. It was the best time of his day. The only thing that would make it better was to meet with some foreign visitors along the way. He would say hello and start conversations. There was plenty to talk about. The scenery was natural and the radio always provided timely topics. This improved his knowledge of the outside world and his English proficiency, to say the least.

Ma Yun had an impressive looking face. His broad smile and eagerness to speak English exuded warmth and hospitality. Most tourists welcomed the bold openness seldom found in this relatively secluded society. Many were quick to establish friendships with him, finding him cheerful and ambitious. Their descriptions of the outside world sparked his curiosity and fascination. These encounters had lasting effects on the formation of his values and respect for different cultures.

Born in 1964 to a modest family in Hangzhou, Ma Yun's life was rather bumpy. After failing the college entrance examination twice, he busily prepared for the next and took odd jobs for added stimulation. On the third time his overall score was still borderline, but his excellent English proficiency stood out. He was finally admitted to the Hangzhou Normal College as an English major. To his luck and alma mater, Ma Yun was forever grateful.

During this period, a friend he made in West Lake invited him to Australia as a houseguest. This trip was a real eye-opener. He had been under the impression that life in socialist China was paradise, and that capitalist West was "in hot water", yearning for salvation. After a month's stay in Australia, the question of how to help China out of poverty became deeply rooted in his mind.[37]

Dedication to SMEs

Ma Yun graduated from Hangzhou Normal College in 1988 and was assigned a teaching position at the Hangzhou Electronic Technical Institute with a five-year commitment. He took a keen interest in the world around him. The City of Hangzhou and the Province of Zhejiang were historically famous for entrepreneurship. Small and medium-sized enterprises (SMEs) were everywhere and became the main contributors to the economy after agriculture. Taking note of their need for English translation to pursue export commerce, he opened a language translation service. He also made available the premise to exhibit seasonal goods from SMEs, which broadened his understanding of their plight.

In 1995, Ma Yun made his first trip to the U.S. In the Seattle area he was introduced to the internet and was astounded by its immense capability. He immediately envisioned the internet as a powerful tool for SMEs in China and a technology that could change their life. When he returned, he raised RMB20,000 (approximately US$2,000) to found a company called The China Yellow Page. It was meant to be a portal to information that could facilitate SME business transactions.

The company was quickly acquired by state-owned Hangzhou Telecom as a joint venture (JV). His meager investment was revaluated to increase 30 times and he would own 30% of the JV. However, he soon became unhappy with the management, which steered the company away from SMEs. Thus, he sold his shares at a discount and left the company. He travelled to Beijing where his talent was quickly noted. The Ministry of Foreign Trade asked him to set up websites and prepare electronic exhibits. Again, he felt that the leadership circle's attention was bent toward multinational corporations (MNCs) and state-owned enterprises (SOEs), where big money spoke. With more experience and a better understanding of international trade, he felt it was time to pursue a business career of his own.

A Star Is Born

On February 21, 1999, at home in Hangzhou, Ma Yun met with a group of his closest friends and briefed them on his vision and plan for establishing an internet company. He requested their participation. Including Ma Yun and his wife, Zhang Yin, there were 18 co-founders, later fondly referred to as the "18 Lohans" (十八羅漢), as in Buddhist culture. The total cash investment for the group was RMB500,000. The company, called Alibaba On Line, was designed to be a B2B platform for SMEs to conduct trade over the internet. The name "Alibaba" was chosen because it was similar to a name in the widely read story "One Thousand and One Nights," catchy and easy to remember.

In the new millennium, global trade increased rapidly. However, for small producers in China, finding outside buyers was difficult due to their lack of experience and expertise. Those in the MNC supply network would have to pay handsome commissions. For small foreign buyers, finding reliable sourcing in China was equally difficult because of the relatively closed society and limited information available. Thus, this new B2B platform came at just the right time. Its motto was, "There is no business that is difficult to do."

Both producers and buyers could log on to the platform and follow up with any contacts as they pleased. There was no charge to the buyers or producers. However, a producer could become a member of the website voluntarily by paying a token membership fee. Members received priority in getting to the front of the supplier selection list. They were also rated depending on the number of times they had successfully consummated deals. This was a certification of quality that provided tangible value to producers. Most of the revenue was from advertisements, however. The novelty of this platform was greeted with enthusiasm by SMEs.

At this point, internet technology was still new in China. It took time for SMEs to become familiar with the operations. Between 1999 and 2000, the business was slow, the start-up fund was drying up quickly and the company was struggling to survive.

It was fortunate that investments from Soft Bank, Goldman Sachs, Fidelity and other institutions totaling US$25 million were settled. These investments arrived at a critical moment for another reason: the internet bubble soon collapsed and many technology start-ups suffered premature death.

In December 2002, Alibaba reached the breakeven point with 1 million memberships. After this milestone, business took off at a phenomenal rate. It was further propelled by China joining the World Trade Organization (WTO), an event that Ma Yun had anticipated while he was planning for the company.

Shopping Made Easy

In May 2003, Taobao.com was founded as a customer to customer (C2C) online shopping platform. Unlike the initial B2B platform, which focused on serving businesses, Taobao.com offered a variety of products for retail sale to consumers.

Again, registration was free and there was no commission fee. This platform was extremely popular and the company's revenue rapidly surged, reaching its goal of US$1 million per day. At the beginning, eBay, the American internet company, dominated the field of online shopping. It exerted its influence with local telecom providers to keep out any advertisements from Alibaba. This coincided with the SARS outbreak and people travel was restricted. To counter this double isolation, Alibaba posted Taobao's ads in full Chinese and on all Hangzhou city buses for millions of passengers to see. The effect was tremendous! Once it took hold in Hangzhou, it spread to other cities almost instantaneously, even without help from the telecom companies. Eventually, eBay retreated to the back row. Ma Yun was quoted as saying, "eBay may be a shark in the ocean, but I am a crocodile in the Yangtze River. If we fight in the ocean, we lose; but if we fight in the river, we win."[139]

In December 2004, Alipay, which started earlier as a third-party payment platform on Taobao.com, became a separate business. This was extremely important in China, as the post-1980s baby boomers became active consumers. This group had purchasing power and preferred to do quick and easy deals. They generally did not carry checkbooks and some did not even have credit cards. Alipay allowed them to leapfrog into e-commerce.

Expanding Technology's Reach

With these three business platforms on the market, Ma Yun felt Alibaba needed a good search engine with good domain recognition. Thus, in October 2005, it took over the operation of China Yahoo! and became a strategic partner of Yahoo! Inc. Ma Yun's relationship with its founder, Jerry Yang, started much earlier. It was reported that Ma Yun was offered to head Yahoo! China while he was planning the Alibaba venture, but that he opted to be an entrepreneur on his own.

In November 2007, Alibaba was successfully listed on the Hong Kong Stock Exchange. This enhanced Alibaba's appeal to the Chinese communities of Hong Kong as well as Southeast Asian nations, which commanded immense economic power. It also facilitated trade with producers in mainland China.

Some Like it Hot

In April 2008, Alibaba established Taobao Mall (TMall.com) as a business to customer (B2C) platform. This complemented the Taobao C2C platform. It offered personal products from global brands to an increasingly affluent Chinese consumer base. TMall became a separate business in June 2011 and had since grown rapidly.

In the Double 11 (November 11) promotion week of 2016, TMall achieved a record of 1 billion transactions per day with total daily values of CNY120 billion, which was approximately a 60% increase overall from 2015.

The Double 11 promotion was joined by other Chinese e-commerce websites, which together stimulated shoppers. Ma Yun explained many times why he had picked e-commerce as his focus instead of mail or gaming. He said that people could get tired of reading mail and playing games, but never of shopping.[141]

The five portals, B2B, C2C, B2C, e-banking and the search engine were the central business units of the Alibaba Group. Thus, the company had established the infrastructure it needed to become a leading global e-commerce organization.

Striking It Big

Starting with 18 employees in 1999, the Alibaba Group had approximately 11,000 employees in 2014 and over 40,000 worldwide in 2016. In a speech in 2008, Ma Yun proudly reported that his company had achieved a daily revenue of CNY100 million in 2002 and a daily profit of CNY100 million in 2004, and that it paid taxes of CNY100 million daily in 2006. In 2013, Alibaba handled CNY1 trillion (US$170 billion) transactions per year, more than eBay and Amazon combined.

Undoubtedly, Alibaba was a phenomenal success, not only in China, but also in the world. It was estimated that by September 2014, Alibaba's websites had accounted for over 60% of the parcels delivered in China and 80% of China's online sales. Ma Yun claimed that the Alibaba Group had created more than 30 million jobs in China.

Internationally, AliExpress.com, which offered products to international online buyers, was the most visited website in Russia.

On September 18, 2014, Alibaba formally made its IPO of US$68 per share in the New York Stock Exchange. At a total market value of US$21.8 billion, it became the biggest IPO in U.S. history.

Ma Yun served as Alibaba Group's CEO for its first 10 years. He stepped down in May 2013 and became Executive Chairman. Jonathan Lu, one of the 18 original Lohans, became the company's CEO. Joseph Tsai, who had been instrumental in early fundraising and in steering the company's finances, was Executive Vice Chairman. In May 2015, Lu was succeeded by Daniel Zhang. In 2016, Alibaba's official website showed that the company had over 46,000 employees worldwide and a revenue of over CNY101 billion (approximately US$18 billion).

In October 2014, Alipay was reorganized to become Ant Financial Services Group and became an affiliate company of the Chinese Alibaba

Group. Ant Financial Services Group was "dedicated to creating an open eco-system, enabling financial institutions and partners to make rapid progress toward internet + goals." It also provided inclusive financial services to small and micro enterprises and individual consumers. This was a much broader scope than that of the original Alipay.

A recent financial report showed that the Alibaba Group, including the Ant Financial Services Group, paid a total of CNY23.8 billion in taxes in China for 2016, up from CNY17.8 billion in 2015 and CNY11 billion in 2014. Thus, Alibaba was the largest taxpayer of all Chinese internet companies.[140]

Preparing for the Future

Since 2010, Alibaba has embarked on an ambitious merger and acquisitions program. It expanded into online music, smartphones, smart remote apps, mobile internet, motion pictures, video programing, car sharing, entertainment gaming, healthcare, newspapers and professional football. This was quite atypical of Alibaba and appeared to suggest that the company was changing focus to broaden its base for future growth.

In celebration of its 10th anniversary, in September 2009 the company established Alibaba Cloud Computing (AliCloud). It aimed to build a cloud-computing infrastructure that covered data mining and processing, and to work with clients to build PaaS and SaaS. Clients were considered as business partners to promote the services that AliCloud co-developed with them. This program was launched due to the recognition that the internet had brought about a rapid change in the business world, and the industry and the entire socioeconomic environment. Facing these changes, Alibaba was no longer just an e-commerce company, but one that would drive changes and strived to stay on top of it. Obviously, this was a vision that required a new business model. It operated R&D centers in Hangzhou, Beijing, Hong Kong, Singapore and Silicon Valley.

Ma Yun decided to turn over company control to others and dedicated himself to corporate direction, overall policy, corporate culture and initiatives related to education and public good. He prepared the company for the future, as cloud computing and big data applications were going to be the underlying factors of all walks of life, just as the internet had become a global agent of change 10 years before.

In October 2016, before the AliCloud Computing Conference in Hangzhou, Ma Yun said, "We anticipate the birth of a re-imagined retail industry driven by the integration of online, offline, logistics and data across a single value chain. He continued, "This is why are adapting, and it's why we strive to play a major role in the advancement of this new economic environment." With this new vision, his mind was still on SMEs. However, this time, he was concerned with SMEs not only in China, but also around the world.

He said, "We believe the commerce infrastructure we have created in China — marketplaces, payments, logistics, cloud computing and big data, all working in concert — can be applied on a global scale to lift up small and medium-sized businesses and ordinary consumers around the world." He also summed up his view in a letter to the annual shareholders' meeting.

Legacy

The rise of the Alibaba Group over 18 years was one of the most fascinating stories in the history of innovation and entrepreneurship. Analyzing its trajectory, some characteristics can be highlighted.

- **Mission**

  Ma Yun had shown a strong sense of social responsibility since he was a young man. His observations in Australia and the U.S. fostered his vision and mission. He wanted to help China by teaching SMEs to use the internet. Noting that SMEs provided 70% of the jobs in China, by making them stronger and more profitable, he was going to help build a better and more affluent society. As General Electric had brightened the world with electric lights and Walt Disney had made people smile with entertainment, he wanted to empower SMEs to do better business by using the internet.

  This was the mission he gave himself and his colleagues. He made it a company promise to assure SMEs that there was no business that could not be done. By mapping company strategies, setting policies and training incoming employees, he conveyed this mission to everybody in the company. Later, when Alibaba expanded to other countries, he wanted the company to help the SMEs in those countries. He said Alibaba was successful only if its customers were successful.

- **Strategy**

Ma Yun was skeptical of the national focus on manufacturing. Why? He said China lacked energy, materials and know-how. Furthermore, labor costs were going to increase and pollution was going to worsen. He felt that e-commerce was better for China and that the internet could drive economic growth. Yes, China had only a brief history of internet use. So did almost everyone else. This gave Alibaba the opportunity to make history. China's joining the WTO was going to greatly enhance the demand for e-commerce. He believed Alibaba should be prepared for it.

His first strategy was to build basic e-commerce infrastructure in China. This was why he established the B2B platform for producers and buyers when the company started. To attract usage, he made transactions on the platform free of charge. Doing business online gave producers a 30% added profit margin. As SMEs became familiar with the internet, Ma Yun introduced the next platform — the C2C model, followed by Alipay payment platform. This strategy was in the e-banking realm. Next, he merged with Yahoo! China to build stronger internet resources for the company's users. Finally, the B2C platform (TMall), which was an upgrade of Taobao, addressed the needs of the fast-growing affluent customer base.

By progressing from one platform to another, Alibaba built the infrastructure required for e-commerce. As Ma Yun liked to characterize it, this was Eastern wisdom coupled with Western management aiming at the global market.[141]

When all of this system's components fell into place, cloud computing and big data applications emerged from the scene. This ushered in the ambitious AliCloud.

In a nutshell, the Alibaba Group under Ma Yun's leadership put vision into action with a strategic play out.

- **Values**

Ma Yun recognized that for a fast-growing company with new recruits from different academic, business and social backgrounds, it was essential for them to understand its mission and to share its values. Without this, the company was only an aggregate of people who would be in each other's way creating friction. Thus, he repeatedly lectured on the values of Alibaba.

The companies first advocated nine corporate values, but later narrowed them to six. Ma Yun liked to compare the six points to the invincible *Holy Sword of Six Nerves* (六脈神劍) of his favorite legend *Fong Qing Yang* (風清揚) from the famous author Jing Yung (金庸).

* ***Customer First:*** The company achieved success only if its customers were successful.
* ***Team Cooperation:*** The most valued asset was people who could also work in teams.
* ***Embrace Change:*** Change was inevitable; leaders grew from change.
* ***Integrity and Trust:*** The last line of defense that was not to be treaded upon.
* ***Passion:*** Total dedication to achieving goals.
* ***Industriousness:*** Work hard and work smart.

Ma Yun stressed that Alibaba was a service company, not a technology company. These values were necessary for everyone to understand and use to guide their behavior. Only with this behavior could a company of common people make uncommon achievements.

- **Teamwork with global perspective**

  One distinct characteristic of Alibaba was its global perspective. Ma Yun's early trips to Australia and Seattle were defining moments in his life. He made friends and learned from travel. It is of interest to examine Alibaba's top management team in its formative years, which some referred to as the "Triangle."

  — Chief Operating Officer (COO): Ming Seng Guan, a British national from Hong Kong with over 20 years of experience at General Electric.
  — Chief Financial Officer (CFO): Joseph Tsai, a Canadian from Taiwan, trained in law and finance, joined Alibaba from INVESTOR AB in Sweden.
  — Chief Technology Officer (CTO): Jun Wu, a U.S. citizen from Shanghai, was Vice President of Yahoo!.

The COO, CFO and CTO all reported to the CEO, Ma Yun, who was a Chinese entrepreneur from Hangzhou.

The participation of outside talents in the highest level of corporate management was extraordinary. Thus, the company could naturally see things from different perspectives and had a good understanding of global developments in its field. Ma Yun valued such global talents and was able to win their trust.

- **Execution**

  The company swiftly transformed ideas into action. Western pundits speculated why the Chinese government had given Alibaba so much freedom in establishing an internet company, e-commerce platforms and an e-banking business, all highly sensitive fields. Has it been granted preferential status by the ministries or special policy measures by the state? Was it due to the participation of some "revolutionary nobles"? Considering the background of the 18 Lohans and the top management triangle, the answer was clearly "no."

  Alibaba was a doer's company. It gained commanding status in all of these fields because the fields were wide open at the time. When the policymakers and regulators realized this, Alibaba was already up and running, and it ran faster and better than other foreign players.

  Furthermore, Alibaba never asked any government for subsidy or any bank for loans; it did it on its own. In fact, it provided quality jobs to the community and paid a great deal of taxes to various levels of government.

  The company committed to its mission, shared its values and mapped out its strategy and goals. With these elements in place, the talented team worked swiftly. It was a textbook case in play.

Looking Onward

On many occasions, Ma Yun expressed his wish for Alibaba to exist for 102 years, so that it could hit a record of three centuries between 1999 and 2101.

Change was inevitable. It is hard to predict the evolution of a big global company. Not for 100 years, nor 50 years. In Chapter 15 of this book I discussed two unfortunate examples in the automotive industry that have been very much in the headlines over the last few years: General Motors and Volkswagen.

There were many similar stories of management breakdown of corporate giants in other manufacturing industries. Are service companies such as Alibaba immune to defective products and processes? Not likely.

When I was a graduate student, my professor warned about the perils of computers. He said, "When a calculator makes an error, it is a mistake; when a computer makes an error, it is a disaster." I can add, "When the internet makes an error, it is a catastrophe," as the mistake can travel to all corners of the world within seconds.

Accessibility versus Liability

The crown jewels of the Alibaba Group, Taobao.com and TMall.com, handled tens of millions of products daily. This was certainly good business. However, it was not a business without risk or liability. Any substantial problems in product or in transaction could tarnish the platform's image, to say the least. Such problems might not have been Alibaba's own making. The platforms were easily accessible by all kinds of companies around the world. Monitoring their authenticity was a horrendous challenge. Recently, complaints were lodged in the U.S. against the Alibaba Group for allowing the circulation of counterfeit products.[141]

Competence versus Complacency

To err was human. As time passed, attention, burden and metabolism changed people, increasing the likelihood of error. The same was true for corporations. Competency gave way to complacency, which led to arrogance and ignorance. Laozi said, "If one cautions about the ending as the beginning, there will be no failure" (慎終如始, 則無敗事). He said this because he knew it was seldom done. I surmised that this was an important factor in the sad stories of General Motors of U.S., Volkswagen of Germany and more recently, Kobe Steel of Japan (see Chapter 14, "Managing Change").

Focus versus Diversification

When Alibaba started, it focused on SMEs. All of its programs and attention were on the dynamics and needs of SMEs. This vision stimulated the company's success. The company's portfolio now showed a wide spectrum of businesses. Certainly, in addition to having a broader economic revenue base, the Alibaba Group could have a closer feel of how the social environment would change. However, the battlefield was now much more extended and the focus much more diffused. Was the company's orginal vision being upheld? Is the company structured or organized to master this level of diversification? This is a question that the executives of Alibaba group should examine cautiously.

Globalization versus Localization

Balancing globalization and localization was challenging not only in China, but also in every country the company did business with. Each country had a set of boundary conditions that the company had to live with. In a way, the constituency of Alibaba was like that of the United Nations in its complexity.

Agility versus Stability

Alibaba's leadership in e-commerce was, to a large degree, due to its agility. It quickly recognized problems or opportunities and took swift action to embrace them. This reflected its values. Once a change was made, it needed continuity and stability to build the momentum to achieve results. This was akin to walking on a tightrope and was always the management of changes and risks.When management got increasingly fatigued, the risk would invariably increase.

Management of Change

Change was a fact of life. It could not be avoided, but it could be managed. Laozi observed the nature and outlined its rule. He said, "Fortune breeds misfortune; misfortune leads to fortune" (福兮禍所伏, 禍兮福所倚). Thus, change was cyclical. In good times, one needed to maintain humility and preparedness. In bad times, one needed to be assertive and vigilant. Knowing the principle was not enough. A corporation needed to institute the mechanism for action, review and renovation, just like any product or service it administered. This would reflect on how well a company nurtured its soft power for growth and sustainability, on the first day and every day.

Epilogue

On February 2, 2017, the University of New Castle, Australia, announced that Chinese billionaire Ma Yun planned to donate US$20 million to fund a scholarship at the university to honor his late friend Ken Morley.[142] According to the news report, Morley welcomed young Ma Yun to practice English in the early 1980s in Hangzhou, invited him to visit Australia and remained close friends with him afterward. Ma Yun commented on the occasion: "The scholarship is to honor the experience and the special relationship I formed with the Morley family." He observed that the scholarship program hoped to "inspire, educate and cultivate tomorrow's leaders. Alibaba was built by young people and we are committed to lifting up and empowering students so they can reach their dreams and ambitions." This was a very fitting description of the entrepreneurial spirit.

Chapter 26

TENCENT HOLDINGS LIMITED

The Growth of an Internet Giant

Tencent was once considered the best-kept secret of Shenzhen. It was a company not related to any SOEs or funded by government grants. It was one of the rare species in China: a technology giant that rose to fame entirely on the efforts of young entrepreneurs who had no prior experience outside of China.

The internet company had five co-founders: four graduates of Shenzhen University and a young engineer from Xi'an who returned to Shenzhen. They were struggling young college graduates at the time the company was established in 1998. 18 years later, the company was second only to Facebook in its number of users worldwide, but remained number one in group chat users (856 million versus 800 million). Ma Huateng (also known as Pony Ma), leader of the founding group, had been CEO since its establishment.

I will focus on the early part of the company's history with two questions in mind: what brought these young engineers together to pursue entrepreneurship in Shenzhen? How did the start-up company strive to overcome the challenges it encountered?

The Founders

The founders of Tencent Holdings Ltd. all belonged to the post-1970s generation, all had family ties in Shenzhen. Four of them were classmates at the same Shenzhen high school and went to Shenzhen University upon graduation. The university was established in 1983. It had a beautiful tree-lined campus in the Nanshan District, which later became a hub of innovation. It deserved more credit as a cradle of entrepreneurship.

Pony Ma (馬化騰), the leader of the group, born in October 1971, was a student of excellent academic ability. When he graduated from high school in 1989, he had hoped and prepared to go to Tsinghua University in Beijing or Fudan University in Shanghai to study astronomy. However, the Tiananmen Incident had just occurred in Beijing and his parents wanted him to stay close to home. Therefore, Shenzhen University was the natural choice for him. With no astronomy department to choose, Pony went into electronic engineering and computer science.

There, he was happy to meet his high school classmates, Zidong Zhang (Tony, 張志東) and Chenye Xu (Daniel, 許晨曄). Yidan Chen (Charles, 陳一丹) was also at the university taking chemistry as a major. They all entered Shenzhen University for the same reason. No wonder university records showed that the entering class of 1989 consisted of students of exceptional quality, much better than any previous year.

Pony demonstrated a special interest and ability in the computer language C and graphic user interface. In his senior year, he received practical training at a local computer company and independently developed a graphic system capable of analyzing the trend of stock transactions. The company liked it and offered him CNY50,000 for the system, an amount equivalent to roughly three years' salary for a graduate in his field. This gave Pony his first lesson on the value of knowledge.

After graduating in 1993, Pony Ma was quickly offered a job in a newly established company, Runxun Telecommunication, in Shenzhen. The company was marketing a device for instant messaging called the "beeper." It was a device clipped on the waistband enabling its user to pick up messages that came through a wireless telephone network, anytime and anywhere.

Pony's job was to develop software that could link the beeper to different signal-emitting systems. He was doing very well and was paid accordingly. He was at the bottom tier of a large bureaucracy and often fondly addressed as "Xiao Ma, or, Kid Horse" which gave rise to the nickname "Pony."

His job had an upside — it gave him the freedom to delve into another hobby: the Fido network, a private network through fixed telephone lines. One could post content through the network, point to point, as long as one could hold on to the telephone line. Later, Pony learned how to accomplish this using a modem. Reaching and talking to wide, and sometimes wild, audiences without face to face contact gave him a great sense of satisfaction. Thus, he had a personal appreciation for the pleasure and power of the network. He gradually built up his "Ponysoft" station, which gained reputation in the Shenzhen area.

Tony Zhang, on the other hand, was well recognized for his achievements in mathematics. He pursued post-graduate studies at South China University of Technology in Guangzhou. Daniel went north to Nanjing University for graduate studies in computer science. Charles took a job at the Shenzhen Imports Inspection and Quarantine Bureau. He also studied for a Master's degree in economic law on the side.

In September 1996, Tony finished his graduate studies and returned to Shenzhen. He took a job in a computer company that provides network services to beeper stations. Very soon, he was reunited with Pony.

The two met up frequently. They talked about old times and the exciting emergence of the internet. The news of Sohoo (later "Sohu"), Sina and NetEase was constantly on their minds. But these products appeared to be overlapping with only minimum difference. What could they do to make the internet reach more people and enrich their lives in more distinctive ways? On one spring day in 1998, Pony said to Tony, "Let us set up a business together." Without hesitation, Tony agreed.

In reminiscing, Tony later said, "We did not think of making money; we just wanted to do something different and exciting and of value to the society."

They quickly found Charles and Daniel, who were both in the Shenzhen area. Daniel was working for the Shenzhen Digital Communication Bureau and eager to help. Charles had been married for two years and had a good, stable job with the Inspection Bureau. Weighing the uncertainty and risk of entrepreneurship, Charles did have a moment of hesitation. However, his wife said, "Don't worry, I have a job too."

Thus, the four old friends joined to make plans for the new company. Very quickly they recognized a serious gap in their skillsets — none of them knew the art of selling products or services.

A fifth member then came onto the scene: Liqing Zeng (Jason, 曾李青). Jason was born in 1970 and was nearly two years older than Pony. He was a graduate of Xi'an Electronic Technology University, where he majored in telecommunications. After graduation, he returned to Shenzhen and worked with Digital Communication Services. He had a bigger build and was a cheerful and eloquent communicator. In many ways, these characteristics were complementary with those of Pony and Tony. Jason was the marketing manager of a JV business of the Telecom Services Company, a SOE. At this point, the business was under pressure to reorganize and he was examining his own career path. He wanted to build his future in the real business world, certainly not a quasi-business unit of a government agency. Therefore, through a mutual friend, the five met and very quickly shared the same vision for the future.

The Company

A simple division of responsibility was outlined: Pony Ma was responsible for strategy and product, Tony Zhang for technology development and Jason Zeng for marketing and sales. This understanding was maintained for many years as the company grew.

They each decided to take their own savings to establish the company. With a total investment of CNY500,000, the minimum investment required under business law, Pony owned 47.5%, Tony 20%, Jason 12.5%, Charles 10% and Daniel 10% of the company. As for the company's name, they agreed on "Tencent". Why? They wanted to take after the U.S. giant Lucent, thus the "cent." And, "Ten" was the only one of several options that was not preregistered. It was also part of Pony's name. In later days when the company's profit hit the million mark, Jason gleefully mentioned that the name served as a reminder that the company grew from selling instant messages at "ten cent" a piece.[143,144]

The company had three full-time employees to start with. Charles and Daniel were advised not to quit their formal jobs as yet. The new company had to run lean. The formal date of registration was November 11, 1998.

All the founders were very young and came from modest backgrounds. There was no specific government financial support or SOE funding for the company. Stimulated by innovators and entrepreneurs in the U.S. and China, they chose entrepreneurship to pursue their vision. They were on their own and confident that they were going to make a difference in society. The founders treated each other as equals and no one claimed special credit, privilege or intellectual property rights.

Time and space were in their favor. First, they arrived at the window of opportunity in the early phase of internet technology. Second, they understood the immense market potential that China presented. Finally, Shenzhen was a fertile ground they were familiar with and could access the resources they needed (see Chapter 18). They recognized this golden opportunity and decided to take action.

Good Product but Dwindling Market

Tencent planned to market a rather unique product. As Pony explained, it sought to develop computer software systems to connect the internet and beeper and thereby allow users to receive news and e-mails, all in Chinese. In other words, without making a telephone call, users could freely receive information. This was essentially a "wireless networking beeper system." Beeper stations were the target customers.

This sounded like a great product concept. In terms of the enabling technology, Pony had five years of experience working with the beeper system and Tony was an expert in system integration. They were confident that they could do it. But, there was only one problem: the usage of beeper was rapidly contracting. Why? The mobile telephone had gained popularity after 1998. For example, Motorola, king of the beeper in China, decided to close down beeper operations by the end of 1998. Worldwide, beeper sales were falling precipitously. The CNY200,000 unit price of the software system Tencent developed dropped below CNY100,000. It soon reduced further to CNY50,000 and CNY30,000, which was basically at cost. To keep the company from going under, Jason busily solicited jobs, such as developing e-mail accounts and designing websites.

Nurturing OICQ

ICQ was another development that occurred concurrently with the beeper. It was a software system that enabled direct news reporting, document transmission and two-way conversations on the internet. The only requirement was that the recipient had to be preregistered and their names listed to facilitate the operation. The ICQ ("I seek you") system was developed in 1996 by three Israeli youth after completing their military services. The company they formed — Mirabilis — was registered as providing instant messaging services. The system was very much welcomed by young people. Within a year, it was rated the most-used instant messaging software in the world. By the end of 1998, when Mirabilis was acquired by America Online (AOL), it had reached 10 million users.

Pony took note of ICQ when Mirabilis started. He submitted a proposal to the Runxun Telecommunication management for a feasibility study on launching such a system in China. The proposal was turned down, as the system's profitability was doubtful. However, other people in the region had also noticed ICQ. An IT company in Taiwan was working on a standard Chinese version called "CICQ." In 1998, a simplified Chinese version was marketed in mainland China as "PICQ." A company in Nanjing also marketed an ICQ-like product. However, market reception was meager at best.

The Tencent co-founders noticed an announcement from the Guangzhou Telecommunication Bureau for an invitation to bid on an ICQ-like tool in Chinese. They promptly submitted an outline for a program called "Open ICQ" (OICQ). It was regretfully but not unexpectedly rejected. This topic was discussed and debated within Tencent. Although the profit model was unclear, Pony decided to nurture OICQ a bit more, under Tony's supervision.

Continuous Improvement of OICQ

Tony Zhang led a small group and took pains to improve the system. The first issue was moving the user list and contents to centralized servers. This was done to consolidate the records of many separated terminals to servers under a unified management system. This situation was unique to China, because at that time most users did not own a computer: they relied on various internet cafés to get online. If the database was in a café's user terminal, one lost the list and content if one moved to another café. This was obviously undesirable. Although consolidation was a costly step, it was necessary and extremely important to promote usage.

The second issue was compressing the software size so that it could be downloaded and processed at a reasonable speed. This maximized the number of users online at any one time.

Other improvements included providing personalized profiles, audible signals and user software upgrades. OICQ was a different entity from the initial ICQ as a result of these small but swift innovations. With these improvements, they were able to support the rapid growth of users in China.

What did competitors do while these activities were taking place at Tencent? Very little. After the US$400 million acquisition, AOL made little conscientious effort to improve the system. The competitor from Taiwan was not a threat due to its lack of knowledge of the conditions on the mainland. Other small players in China were relatively dormant due to the lack of technical support. Tencent announced its first version in February 1999, only three months after being formally established.

After a product's software was posted online, debugging was continuous. There were incidents of rapid replacement, with up to three substitution versions in one week. This reflected the stated corporate guidelines: take small steps, do quick runs and make rapid substitutions.

Toughing It Out

In China, servers were expensive, but labor was cheap. Few companies, especially start-ups, could afford multiple servers because of the cost. Therefore, they had to work with their existing servers to get the most out of them. This was maximization and optimization.

In Tencent's case, the co-founders had invested approximately 40% of its cash into purchasing a very powerful server to be the major workhorse. Knowing this was all they could afford, Tony worked very hard to build a parallel computing architecture to make the most of the server. Other improvements in the algorithm were also made. This enabled Tencent to achieve a niche advantage in computing capability that no others in the field could match. Thus, continual improvement became a core value of Tencent.[145]

The number of users increased exponentially, quadrupling every three months. This was totally unexpected and put unexpected load on the server. In addition to the various improvements being undertaken, the co-founders spent time in electronic outlet stores searching for parts and putting together "make do" versions on their own. Obviously, the functions and quality did not compare to those of authentic servers and generated more bugs. Therefore, they had to maintain on-call service engineers at all times. All of these efforts enabled Tencent to withstand beyond the threshold of 1 million users.

However, the company's finances got increasingly tighter. Pony and Jason had to look for other jobs, such as designing websites and building service providers, so that they could sustain the operation.

Tencent's reputation quickly spread by word of mouth. User polls were very enthusiastic; they helped choose the Penguin profile, which became so popular that it was later turned into a separate business. With the overwhelming enthusiasm and support of Tencent's users, the co-founders felt they could not quit and let the users down, even in such dire financial circumstances. They would just have to tough it out for the users.

Misery Seeks Company

Nine months after establishment, user registration reached the 1 million mark and continued to show a rapidly increasing trend. However, the company's treasure chest was nearly empty. This financial worry was matched by an unexpected legal worry. Tencent received notice from AOL's lawyers claiming it had infringed upon AOL's intellectual property rights and had to stop using the domains OICQ.com and OICQ.net immediately.

The co-founders decided to tackle their financial problem first, as it was an imminent survival issue. A two-pronged approach was taken. First, they agreed to each contribute more funds, doubling the equity to CNY1 million. They also halved their salaries, which were already minimal. Second, they actively sought a buyer to take over. Nearly 10 groups of possible investors were contacted. However, no one was willing to take over an apparently dead-ended company.

Entering Venture Capitalism

Luckily, a venture capitalist, the International Data Group (IDG), appeared on the scene. The IDG was a medium-sized U.S. venture capitalist that had been in Shenzhen for a few years. It was brought to their attention that a small internet start-up had several million users but had no business model and was about to fold. The group showed up to discuss Tencent's business plan. As IDG manager Shu Wang flipped through the pages that Jason put together, he asked, "What do you see as the future of your company?" Pony Ma calmly and cautiously said, "I really don't know." It impressed Wang that the team leader was honest, pragmatic and trustworthy.

In reality, the IDG liked the project because users in China had already shown a preference for it. More importantly, in the U.S., AOL negotiated to buy Mirabilis and was willing to pay a price of US$287 million. There had to be something valuable in Tencent, although a good business model had yet to be found.

Eventually, the IDG partnered with Pacific Century Cyber Works (PCCW) of Hong Kong to buy 40% of Tencent for US$5 million. The agreement stipulated that half the cash was going to come at signing with the remainder injected within one year, as long as the number of users reached the set milestone. Thus, the deal was made. Wang further offered to put in the money from his personal funds to relieve Tencent's financial crisis before all of the documents had gone through legal processes. The agreement was signed in April 2000.

The Bubble Breaks

At this point, the world was about to experience the collapse of the internet bubble. In March 2000, the Nasdaq Index hit 5,000 and the Shanghai Index peaked at 1,700. Shenzhen was about to establish its own stock exchange. The world's stock markets were bullish with internet-related issues.

However, in the second week of April 2000, the Nasdaq Index dropped precipitously by 40%. An estimated US$8,500 billion of market value had essentially evaporated. Famous internet stocks, such as Cisco, Yahoo! and Amazon, all ended in disastrous loss. The price of three Chinese stocks, Sina, Sohu and NetEase, dropped to below US$1.

Luckily, Tencent was able to escape unscathed. Whereas many others were gasping for breath, Tencent gained new strength with the new injection of funds. Between the IDG/PCCW agreement and the burst of the internet bubble, it had only a six-month window. Jason Zeng commented, "Bad times, like cold weather days, are perhaps the best time to incubate, [as] nobody will care to bother you."[146]

QQ Enters the Limelight

On March 21, 2000, a U.S. District Court ruled that the names "OICQ" and "OICQ.com" belonged to AOL. This was not unexpected. After consideration, the co-founders decided to change the domain name to simply "QQ."

In November 2000, a new QQ2000 protocol was published and the domain name www.Tencent.com was highlighted. QQ2000 signified the start of a new chapter in which the instant messaging company would transform into an internet service and a virtual community. This was the basic framework of all future Tencent product development work. All company-related logo images were changed to the Penguin series. Many past logos that were similar to Disneyland personalities, such as Mickey Mouse and Donald Duck, were scrapped. This removed any possible copyright issues with the famous brand.

Restructuring Investors

The IDG suffered from an unfortunate consequence of the great bubble burst. Many of its investments had encountered substantial loss and it wanted to exit from Tencent to consolidate its assets. For Tencent, responses from users were excellent, but it was quickly burning money to meet its server demands. Thus, another round of fundraising was necessary.

Several major ISPs in the region, such as Sina, Sohu, Yahoo! China and Tom.com, were approached. However, they were skeptical about Tencent because it appeared to have little tangible advantage and the prospect of profitability seemed remote. During this frustrating time, however, the number of registered users kept rising and was about to reach 100 million.

Then luck struck again. It came from the Multichoice International Holdings (MIH) Group in South Africa. The company was fascinated that Tencent users were increasing by approximately 150,000 per day. It realized that this level of accessibility to the vast Chinese population had to have the potential for profit.

The MIH Group offered Tencent US$60 million on two conditions. It wanted to be the major holder, and it wanted to exchange its shares of Vianet.com for Tencent ownership. Pony Ma immediately rejected these conditions. He wanted to maintain control of the company and only wanted cash. Nevertheless, he was pleased with the proposed valuation, which appeared to have increased 12 times in one year.

The IDG and PCCW were overjoyed with the possibility that they could exit. After rounds of negotiation, an agreement was reached in which the IDG transferred 12.8% of the holdings and PCCW all its 20% to the MIH Group. The participation of the MIH Group greatly strengthened Tencent's financial position. Needless to say, PCCW's joy of exit was to be seen as a source of great agony just a few years later!

Around the time of this deal, in May 2001, the Nasdaq Index bottomed out and the market was on an upward trend. Tencent's registered users reached the 100 million milestone in less than two years of establishment.[147]

Business Models

As user registration progressed steadily, the question of how to turn popularity into revenue to sustain the company hung over everybody's heads. They repeatedly debated how the internet could add real value to the user's life in addition to stimulating their curiosity. It also had to find an easy user payment method.

Looking at the experiences of the U.S., Japan and Korea, the co-founders concluded that they needed to generate content and to partner with the network provider. This will also enable them to access payments. Jason Zeng, who had previously worked in the telecommunications industry, was up-to-date with the ongoing industry changes. He knew that the spin-off of China Mobile as an independent business was imminent and began to recognize that a golden opportunity might be in the offing.

The business concept was as follows. Tencent encouraged OICQ/QQ users with mobile phones to register simply by sending an instant message. The users were then given an OICQ/QQ number and could be connected to the network through their mobile phones. Through the network, they could access any content provided and, also, make payments. The income generated by the content was split between the network and Tencent, to their mutual benefit. This was the essence of the "MobileOICQ" promotion on May 17, 2000.

After the China Mobile Communications Corporation was formally established, it set up a mobile internet service, called "Monternet," to manage all of the content businesses with a 30/70 split to the content provider. By the end of year 2000, China Mobile had provided Tencent a revenue of CNY10 million, equivalent to 1 billion sent messages.

Thus, Jason Zeng and his team busily ran around the country with various local mobile providers to set up "MobileQQ" businesses. They installed the servers right inside the providers' sites. By the end of June 2001, when the MIH Group formally took part in Tencent, Pony Ma happily announced that Tencent had reached breakeven for the first time.

By the end of 2001, it had accumulated revenue of approximately CNY50 million and a net profit of over CNY10 million. The Monternet program obviously benefited all of the participating content providers in China. Data showed that Tencent's MobileQQ occupied approximately 70% of Monternet's entire instant messaging volume. With this, Tencent Holdings Ltd. started to take off.

Corporate Governance

As the mobile internet business progressed, the company restructured the management system. It was divided into three groups: Marketing, Research, and Corporate Function. Reporting to Pony Ma, CEO, were Jason Zeng, COO; Tony Zhang, CTO; Charles Chen, CAO; and Daniel Xu, CIO. The five co-founders assumed specific responsibilities within the company.

The co-founders and their first-line executives constituted the top-level management core. They held biweekly meetings to review, discuss and decide on all company-related issues. The meetings started at 10:00 am and lasted until 2:00 am or 3:00 am the next day, when all of the issues on the agenda were resolved or when no participants could continue staying up to talk.

In these meetings, whoever was responsible for a given business area made proposals and took accountability. On strategic matters, the norm was to seek a consensus. Failing to secure a consensus indicated the end of a proposal. Pony Ma, who chaired the meetings, usually just listened. He was not authorized to veto or overstep.

After 2002, Tencent embarked on many ventures to broaden its business base and income streams. These included providing advertisements, business services, instituted memberships, designated naming and numbering, with various degrees of success. In 2003, Tencent started experimenting with news media, games, Q coins and virtual programs, such as Avarta, QQ Play and QQ Show. The programs were very popular and brought in steady income.

As the company activity diversified, it established research centers around the country for technical support. It also worked with universities and technical institutes to upgrade its technological capabilities.

Tencent's relationship with the government was cordial. It never requested any specific policy support. Ying-li Liu, Vice Mayor of Shenzhen in charge of technology and innovation, marveled at what Tencent was able to do on its own initiative and praised it as the city's best-kept secret.[146] I commented to Liu that the environmental factors of the innovation system of Shenzhen were, quite obviously, proven to be conducive to nurturing entrepreneurship (see Figure 9 in Chapter 6). This is a lesson to be learned by many cities including Hong Kong. The government builds the infrastructure, sets the policy, prepares to help but leaves the entrepreneurs alone to do their show!

Launching WeChat

In 2011, Tencent started a cross-platform instant messaging service, WeChat. It soon became the world's largest standalone messaging app, with 700 million active users in China and 70 million outside of China by 2015. It was a major social community network, very much like Facebook in the U.S.[145]

Users in China also tended to use WeChat as a means of ordering goods and services, paying bills and giving gifts. It helped China leap forward into e-commerce, bypassing the steps of writing checks and pressing credit cards. It contributed to establishing a consumer-oriented society in China, changing business norms and peoples' lifestyles. Unquestionably, Wechat is now the signature product of the Tencent Corporation that changes the life style and work style of hundreds of millions of people in the world.

Reflection

The Tencent story is inspirational, if not magical. It involved five young university graduates, rich only in ambition to bring internet technology to enrich the life of people and the resolve to pursue entrepreneurship in an environment of flux and change. With very limited financial resources and external support, they show exceptional perseverance. They are sustained by commitments not to fail the trust of the users and the society at large. They learned to draw on the support of the players of the innovation system and set examples for others.[147] A few points can be highlighted.

- **Win–win Team Spirit with Stakeholders**

  Aligning with a network provider to promote an internet service, MobileQQ proved to be the key factor in Tencent's business revival. This model benefitted both the user and society. It was a win–win policy.

- **Competitive Advantage with Continued Improvement**

  The technical support to move the database from the user's end to the central server was a major step forward. The continual improvements in design and service were characteristic of Tencent's corporate values: quick response, small steps and rapid revision. This resulted in major product innovations.

- **Product Strategy that Aligns with the Cultural Environment**

  Most of Tencent's product development was tailored to China's traditional culture. For example, the Chinese were used to expressing their feelings quietly in writing, rather than openly in words. Thus, short instant messages sent to beepers and through WeChat suited this standard of behavior well. Hence the service spread quickly. It was not clear whether this was a conscious move or a surprise to the company, but the company was quick to take advantage of it. The success of Avarta, Q Show and other virtual reality ventures were based on this strategy.

- **Build Trust Using an Open Leadership Style**

  Pony Ma has been keeping a low profile. He maintained an easygoing and informal leadership style to build trust. There was little formal protocol and people went by their first names. Major decisions were made when consensus was reached to reinforce trust. However, he did not hesitate to make bold moves if needed.

- **Team Spirit**

  The co-founders formed an excellent business team. They relied on their own efforts and cooperation to resolve any problems, instead of going to banks for loans or the government for preferential treatment. Their own efforts afforded them the freedom that the company later enjoyed.

- **Perseverance**

  Tencent had faced continuous financial crisis since its establishment. It needed to maintain customer satisfaction and build participation, which was unmatched by any other players in the field. The co-founders and management team were experienced enough to tough it out at bad times.

- **Search for Quality Excellence**

  The company thrived on its ability to learn and adapt from established technology leads. This earned it the image of an imitator at the initial phase of the company. Although imitation and simulation were normal in the learning process, a technology leader had to develop original technical capacity in the long run. Tencent has now reached the postion of leadership and should strive to expand the frontier of science and nurture future innovations. This was the responsibility of companies that aspire to be the leaders of technology and society.

  In the last decade, Tencent has indeed engaged actively in developing mobile technology, computer software, virtual reality, cloud computing and internet + applications. All are at the frontier of technology.

  As outreach, the company in recent years has shown an increased interest in building the PRD as its stronghold for the future by mobilizing the stakeholders of the innovation system. Pony Ma himself has led the effort to support the Guangdong-Hong Kong-Macau Greater Bay area proposal.[176,177] This action was perhaps taken with broader strategic perspective for the future of the company and the region.

Striving in an Uncertain Future

The rapid growth of Tencent to become one of the world's top internet companies in less than two decades is phenomenal. In June 2017, the market capitalization of Tencent and Alibaba both reached nearly US$350 billion and became larger than any SOE that dominate the Chinese economy. For comparison purpose, the Industrial and Commercial Bank of China was valued at US$260 billion, and China Mobile at US$218 billion.

There are indications that Alibaba and Tencent may now be heading into a phase of possible conflicts with the traditional mainstream industries and the government. Confrontations with the existing brick and mortar industries in China such as wholesale, retail, entertainment, banking, telecommunication, etc., due to different perspectives and approaches are understandable.[172] Conflicts with the government may be more surprising on the surface.

The government has been supportive of the internet industry in the past decades through many favorable policy measures in law, regulation, taxation, etc. Thus the environmental factors of the innovation system as illustrated in Figure 9, Chapter 6 were favorable to the internet industry. But these can change with time as a new socio-economic environment evolves.

In China, the government has been known to pursue control on IT and cyber-security for national security considerations. The internet industry, on the other hand, tends to seek flexibility, mobility, privacy and versatility.[178]

The tension between the state's aspiration for a vibrant and diversified economy and the determination to enhance government's security apparatus is by no means unique to China. As an example, the headline of a front-page report in the *New York Times* on August 10, 2017, says that "Russia Wants Innovations, but Jails Innovators." Obviously, there are vast differences between China, Russia, and indeed, the U.S. Each country has unique boundary conditions confined by different national interests and priorities at a particular time. But the need to seek balance, conciliation and trust among the stakeholders, i.e. the different players of the national innovation system, is of utmost importance for any country. Recognizing that innovation and entrepreneurship will be the deciding drivers for future society, those leaders who understand the essence of building trust and nurturing the spirit of entrepreneurship will reach out and stand up amidst challenges to make proper choices for the long-term growth, harmony and sustainability of their country.

REFERENCES

1. 〈禮記、大學〉: 是故君子先慎乎德。有德此有人, 有人此有土, 有土此有財, 有財此有用。
〈*Li ji-Da xue*〉: "Therefore a king shall strive to set high moral goal. He then can attract people, secure land and natural resources. This will promote the general livelihood." Original Chinese text as follows. 〈禮記、大學〉: 是故君子先慎乎德。有德此有人, 有人此有土, 有土此有財, 有財此有用。
2. 〈孟子 梁惠王 篇〉: 寡人之於國也, 盡心焉耳矣。…察鄰國之政, 無如寡人之用心焉。鄰國之民 不加少, 寡人之民不加多, 何也?
〈*Mencius, King Hui of Liang Chapter*〉: "I have dedicated all my heart to my country. …I don't believe any other neighboring kings paid as much attention to governance as I did. But the population in the neighbor states do not diminish, nor mine show any increases. Why is that?" Original Chinese text as follows. 〈孟子梁惠王篇〉: 寡人之於國也, 盡心焉耳矣。…察鄰國之政, 無如寡人之用心焉。鄰國之民 不加少, 寡人之民不加多, 何也?
3. Wikipedia, "Industrial Revolution," https://en.wikipedia.org/wiki/Industrial_Revolution (Accessed on 08/08/2017).
4. Paul A. Samuelson and W. D. Nordhaus, *General Economics* (New York: McGraw-Hill, 2004).
5. Peter F. Drucker, *Managing for Results* (New York: Harper & Row, 1964).
6. Lester Thurow, *Head to Head: The Coming Economic Battle among Japan, Europe and America* (New York: William Morrow & Co, 1992).
7. United Nations Framework Convention on Climate Change, UNFCCC, 2011.
8. Walter W., Powell and Kaisa Snellman, "The Knowledge Economy," *Annual Review of Sociology* 30, (2004): 199–220.
9. http://www.nber/papers/w15645.

10. Thomas L. Friedman, *The World is Flat: A Brief History of the Twenty First Century* (New York: Farrar, Strauss and Giroux, 2005).
11. Peter F. Drucker, *Innovation and Entrepreneurship* (New York: Harper and Row, 1985).
12. Michael E. Porter, "Clusters and the New Economics," *Harvard Business Review*, December (1998): 77–90.
13. Walter Isaacson, *Steve Jobs* (New York: Simon & Schuster, 2011).
14. Wikipedia, "Global Warming," https://en.m.wikipedia.org/wiki/global_warming (Accessed on 07/08/2017).
15. IPCC, AR5, WG1, "The Physical Science Basis-Summary for Policy Makers, Observed Change in the Climate System," Climate Change 2013.
16. "Wall Street Protesters Are In for the Long Haul," *Bloomberg Businessweek*, October 3, 2011.
17. Adam Karla, "Occupy Wall Street Go Global," *The Washington Post*, October 15, 2011.
18. Occupy Cafe, http://www.occupycafe.org (Accessed on 08/08/2017).
19. Oxfam International, http://www.oxfam.org (Accessed on 08/08/2017).
20. Yeung, Eric and Lin, Otto C C, "Getting Rich in Knowledge Economy." To be published. (2017).
21. Bhasin, Kim, "This is the difference between Invention and Innovation," *Business Insider*, April 2, 2012.
22. J. A. Schumpeter, *Competition, Socialism and Democracy* (London: Rutledge, 1943).
23. Organization for Economic Cooperation and Development (OECD), "Growth in Services: Fostering Employment, Productivity and Innovation: Meeting of the OECD Council at Ministerial Level, 2005," http://www.oecd.org/general/34749412.pdf.
24. Peter F. Drucker, "The Disruptive Innovations," *Harvard Business Review*, August (2002).
25. Robert G. Copper, "New Products: What distinguishes the Winners?" *Research-Technology Management*, Nov–Dec (1990): 27–31.
26. Richard B. Chase and Robert Hayes, "Beefing Up Operations in Service Firms," *Sloan Management Review*, Fall (1991): 15–16.
27. James B.Quinn, Jordon J. Baruch and Karen Anne Zien, "Software Based Innovations," *Sloan Management Review*, Summer (1996): 11–24.
28. Robert B. Rowen, "Software Project Management," *IEEE Transactions on Engineering Management* 37, No.1, February (1990): 10–21.
29. "Cloud Computing: Clashes of the Cloud," *The Economist*, October 15, 2009.
30. Harry W. Cooper, "Programmed Innovations: Strategy for Success," *Research-Technology Management*, Nov–Dec (1986): 12–17.

31. Jonathan A. Obarand Steve Wildman (2015), "Social Media Definition and Governance Challenge," Telecommunication Policy 39, No. 9 (2015): 745–750.
32. G. George and A. J. Bock, *Models of Opportunity: How Entrepreneurs Design Firms to Achieve the Unexpected* (Cambridge University Press, 2012).
33. Rachel Botsman and Roo Rogers, "What's Mine is Yours: The Rise of Collaborative Consumption," *Harper Business*, September 14, 2010.
34. Clayton M. Christensen and Michael Overdorf, "Meeting the Challenge of Disruptive Change," *Harvard Business Review*, March–April (2000): 67–76.
35. David McCullogh, *The Wright Brothers* (New York: Simon Schuster, 2015).
36. U. S. Patent 821393, to Orville Wright and Wilbur Wright, for "new and useful improvements in Flying Machines."
37. Chen Wei, 2015. "This is Ma Yuen," Zhi-Jiang People's Publisher. Original Chinese text as follows. 陳偉, 2015, "這就是馬云, 浙江人民出版社.
38. M. M. Al-Debei, R. El-Haddadeh and D. Avison, "Defining the business model in the new world of digital business," in Proceedings of the American Conference on Information Systems (AMCIS) Vol. 2008: 1–11.
39. Morris Chang, *An Autobiography-Morris Chang* (Taipei: Commonwealth Publishing, 2001). Text in Chinese.
40. Common-Wealth, 2016-11-01, *How did Morris Chang and TSMC maintain the Lead* (Taipei: Common Wealth Publisher, 2016). Text in Chinese.
41. Linda Pinson, *Anatomy of a Business Plan: A Step by Step Guide to Building a Business and Securing Your Company's Future*, 6th Edition (Chicago: Dearborn Trade, 2004).
42. Frederick Betz, *Managing Technological Innovation: Competitive Advantage from Change*, 2nd Edition (New York: John Wiley & Sons, Inc., 2003).
43. The Nobel Foundation, 2010. The Nobel Prize in Physics 2010 — Andre Geim and Konstantin Novoselov.
44. H. W. Kroto, J. R. Heath, S. C. O'Brien, R. F. Curl, R. E. Smalley, "C60: Buckminsterfullerene," *Nature* 318, No. 6042 (1985): 162–163.
45. J. G. Bednorz and K. A. Muller, "Possible High Tc Superconductivity in the Ba-La-Co-O System," *Zeitschrift fur Physik B* 64, No. 2 (1986): 189–193.
46. Lan Xue, Zeng Liang, S.L. Liu, R. P. Mo, "A Report on Innovation Policy of China," (2011). Text in Chinese.
47. Microsoft Accelerator-Beijing, (2016).
48. ET News, "浩鼎案涉貪起訴首開庭! 翁啟惠: 我沒忘記回國初衷" http://www.ettoday.net/news/20170215/866950.htm (Accessed on 08/10/2017).
49. Joel Shurkin, *Broken Genius: The Rise and Fall of William Shockley, Creator of the Electronic Age* (London: MacMillian, 2006).
50. "The Meaning of Bill Gates: As his Reign at Microsoft comes to an end, so does the era he dominated," *The Economist*, June 28, 2008.

51. Richard R. Nelson, *National Innovation Systems: A Comparative Analysis* (Oxford University Press, 1993).
52. B-A. Lundvall, *National System of Innovation: Toward A System of Innovation and International Learning* (London: Printer, 1992).
53. C. Freeman, "The National System of Innovation in Historical Perspective", *Cambridge Journal of Economics*, No. 19 (1955): 5–24.
54. OECD, *Managing National Innovation Systems* (Paris: OCED Publication, 1999).
55. M. Eissebith, "Bridging Scales in Innovation Policies: How to Link Regional, National, and International Systems," *European Planning Studies* 15, No. 2 (2007).
56. Otto C. C. Lin, (1998). "Science and Technology Policy and Its Influence on Economic Development of Taiwan," in Henry S. Rowen, Editor, *Behind East Asian Growth: The Political and Social Foundations of Prosperity* (London and New York: Rutledge, 1998).
57. The Salzburg Global Seminar, Session 441-From Lab to Market: Accelerating Innovation through University, Business and Government Partnership, http://www.salzburgglobal.org/justice/2000-2009/2007/session-441.html (Accessed on 07/08/2017).

    Session Chair: Otto Lin (HKUST), Faculty: Jean-Noel Durvy (EC), C. Judson King (UC-Berkeley), Florian Lennart (LES-UK), Reza Mahdavi (Cisco), R. Polk Wagner (U Penn), Tazio Yakushiji (Japan), Philip Yeo (EDB-Singapore), Lesa Mitchell (Kauffman Foundation), Geza Egyed (Hungary), Bernhard Wieder (U-Salzburg).
58. The Salzburg Global Seminar, Session 451-Innovation in Knowledge Based Economies: Accelerating the Benefits. http://www.salzburgglobal.org/justice/2000–2009/2008/session-451.html (Accessed on 07/08/2017).

    Session Chair: Otto Lin (HKUST), Faculty: Tom Boland (CE-Education, Ireland), Tom Hockaday (Oxford-UK), C. Judson King (UC-Berkeley), Song-de Ma (MOST-China), Guaning Su (NTU-Singapore), Key-Hyup Kim (KITECH-Korea), Martin Kenney (UC-Davis), Lesa Mitchell (Kauffman Foundation), Martin Cronin (CEO-Fortas, Ireland), Jean-Noel Durvy (EC), Andrea Cuomo (ST Microelectronics), Wilfried Wieden (U-Salzburg).
59. Henry Erzkowitz, *Triple Helix: University, Industry and Government in Action* (London: Rutledge, 2008).
60. NASA, "Sputnik 1: Trajectory Details," National Space Science Data Center, January 8, 2017.
61. Wikipedia, "North Korea" https://en.m.wikipedia.org/wiki/North_korea (Accessed on 07/08/2017).
62. Otto C. C. Lin, *The Technology Innovation Quartet: Secrets of Successful Entrepreneurship* (Taipei: Commerce-Culture Publisher, 2013), 48–51.

Original text in Chinese as follows. 林垂宙:"科技創新四重奏: 成功創業故事解密" 台北, 商訊文化出版社, 2013. 簡体字中文版"創新四重奏: 從实驗室到市場"上海交通大學出版社, 2014.

63. Otto C. C. Lin, "Innovation Systems in the Greater China Region," in European Council of Academies of Science and Engineering-2014 Annual Conference (Brussels, 2014). http://www.euro-case.org/images/stories/pdf/annual-conf/Euro-CASE_Conference_Brochure2014_Lin.pdf
64. John A. Matthews and D-S. Cho, *Tiger Technologies: The Creation of Semiconductor Industry in East Asia* (Cambridge University Press, 2008).
65. John A. Douglass, C. Judson King and Irwin Fuller, *Globalization's Muse: Universities and Higher Education in a Changing World* (Berkeley Policy Press, 2009).
66. D. Breznits, *Innovation and the States: Political Choice and Strategies for Growth in Israel, Taiwan and Ireland* (New Haven: Yale University Press, 2007).
67. The original Chinese text is as follows. 〈孔子家訓〉子路曰: 南山有竹, 不鞣自直, 以之為射, 達於犀革! 何學之為? 子曰: 括而羽之, 鏃而厲之, 其入不亦深乎! 子路曰: 謹受教。
68. Wikipedia, "William Shockley" https://en.m.wikipedia.org/wiki/william_shockley (Accessed on 06/09/2017).
69. http://topuniversityranking.com.
70. Michael E. Porter, *Competitive Strategy: Techniques for Analyzing Industries and Competitors* (New York: The Free Press, 1980).
71. "Eastman Kodak Files for Bankruptcy", *The New York Times*, July 19, 2012.
72. Peter F. Drucker, *Management Challenges for the 21st Century* (New York: Harper Business, 1999).
73. Wikipedia, "Cisco System" https://en.m.wikipedia.org/wiki/Cisco_System (Accessed on 06/09/2017).
74. Alistair Barr and Rolfe Winkler, "Google Creates Parent Company Called Alphabet in Restructuring," *The Wall Street Journal*, August 10, 2015.
75. Neil Irwin and Ylan Q. Mui, "Washington Post sales: Details of Bezos deal," *The Washington Post*, August 5, 2013.
76. Jonathan Vanian, "How Hewlett-Packard Plans to Split in Two," *Fortune*, July 1, 2015.
77. John F. Kennedy, "Special Message to Congress on Urgent National Needs" (Boston, MA: John F. Kennedy Presidential Library and Museum, May 25m 1961). Digital Identifier: TNC-200-2.
78. Yeh Yon-Li, *A Biography-Qian Xue-sen* (Shanghai Jiaotung University Press, 2012). The original Chinese text is as follows. 葉永烈著: (2012) "錢學森傳", 上海交通大學出版社。

79. Chang Chieh Hang, Low Teck Seng and Raj Thampuran, *The Singapore Research Story* (Singapore: World Scientific, 2016).
80. Linsu Kim, *Imitation to Innovation: The Dynamics of Korea's Technological Learning* (Boston: Harvard Business School Press, 1997).

81a. Otto C. C. Lin, "Integrating Science and Technology Policy to Promote Economic Growth," in Charles H. C. Kao, and Joseph S. Lee, Editors, *The Taiwan Experience: Past and Future* (Taipei: Commonwealth Publishing, 1995). 林垂宙, "整合科技政策再出發"。高希鈞、李誠主編: "台灣經驗再定位" 第八章。 天下文化出版, 台北, 1995.

81b. Otto C. C. Lin, "Strategy for Industry Upgrade through Technology," *The Commercial Times*, November 20 and 21, 1989, Taipei. Original text in Chinese as follows. 林垂宙,"我國工業科技化之途徑"。工商時報, 1989 年 11 月 20, 21 日, 台北。

81c. *United Daily News*, January 17, 1994, Taipei. Also, *Commercial Times*, January 17, 1994, Taipei. Both reported the visit of the delegation led by Zhao Guan-chao, President of CAS. Text in Chinese.

82a. Otto C. C. Lin, Kang. L. Wang and K. C. Chan, "The HKUST Forum on the Future Development of Hong Kong," May–July, 2002.

82b. Otto C. C. Lin, "Invest in People, Pearl River Delta," in Louisa Tam, Editor, *The Confluence of Affluence: The Pearl River Delta Story*. (Hong Kong: SCMP Books, 2005).

82c. Otto C. C. Lin, "Hong Kong's R&D Growth is limited by its own ambition," *South China Morning Post*, January 5, 2012, Hong Kong.

83. Dan Breznits and Michael Murphree, *Run of the Red Queen: Government, Innovation, Globalization and Economic Growth in China* (New Haven: Yale University Press, 2011).
84. Shirley Zhao, "Caught in the Loop," *South China Morning Post*, February 3, 2017.
85. Shai Oster, "President Xi's Anti-Corruption Campaign Biggest since Mao," *Bloomberg*, March 4, 2014.
86. "Our Bulldozers, Our Rules: China's Foreign Policy Could Reshape a Good Part of the World Economy," *The Economist*, July 2, 2016.
87. The National Academies of Sciences, Engineering, and Medicine http://www.nationalacademies.org/nrc (Accessed on 06/08/2017).
88. U.S. Department of Energy, Office of Science http://www.science.energy.gov/laboratories (Accessed on 06/08/2017).
89. IVA, http://www.iva.se (Accessed on 08/08/2017).
90. Fraunhofer, http://www.fraunhofer.de (Accessed on 08/08/2017).
91. This is the report entitled "Report to the President on Ensuing American Leadership in Advanced Manufacturing" from President's Office of the

Advisors on Science and Technology to President Barack Obama, issued on June 2011. https://obamawhitehouse.archives.gov/sites/default/files/microsites/ostp/pcast-advanced-manufacturing-june2011.pdf (Accessed on 08/08/2017).

92. Industrial Technology Research Institute, http://www.itri.org.tw (Accessed on 08/08/2017).
93. 〈老子〉 第十一章: 三十輻,共一轂,當其無,有車之用;埏埴以為器,當其無,角器之用。鑿戶庸以為室,當其無,有室之用。故有之以為利,無之以為用。
94. Andrew S. Grove, *Only the Paranoid Survive* (New York: Currency/Double Day, 1996).
95. Amazon Web Services, https://aws.amazon.com/What- is- cloud-computing/ (March 19, 2013).
96. 〈老子〉 第四十章: 反者道之動,弱者道之用。天下萬物生于有,有生于無。
97. Terrence Fong, Clayton Kunz, Laura M. Hiatt, Magda Bugajska, "The Human-Robotic Interaction Operating System," Human Robot Interaction (2006).
98. Wikipedia, "Robert S Langer" https;//en.m.wikipedia.org/wiki/Robert_S_Langer (Accessed on 09/08/2017).
99. Rich McCormick, "Donald Trump Says Facebook and Twitter Helped Him Win," TheVerge.com, November 13, 2016. Also, CBS, 60 Minutes, Sunday, November 13, 2016.
100. The original Chinese text is as follows. 〈論語〉顏淵篇: 子貢問政。子曰:足食、足兵、民信之矣。子貢曰: 必不得已而去、於斯三者何先? 子曰:去兵 。子貢曰: 必不得已而去, 於斯二者何先? 曰: 去食。自古皆有死, 民無信不立。
101. *Sen zi- The Art of War*, Leadership: Knowledge, Trust, Mercy, Bravery, Discipline. The original Chinese text is as follows. 〈孫子兵法〉: 為將之道: 智、信、仁、勇、嚴。
102. Wikipedia, "General Motors ignition switch recalls" https://en.wikipedia.org/wiki/General_Motors_ignition_switch_recalls (Accessed on 10/08/2017).
103. Wikipedia, "Volkswagon emissions scandal" https://en.m.wikipedia.org/wiki/Volkswagon_emissions_scandal (Accessed on 10/08/2017).
104. Jack Ewing and Neal E. Boudette, "Guilty Plea for VW and Turmoil May Deepen," *New York Times*, March 11, 2017.
105. The original text of Confucius' teaching regarding "Education of the Individual" is as follows. 致知在格物, 物格而后知至,知至而后意誠, 意誠而后心正, 心正而后身修。
106. The original Chinese text is as follows. 〈中庸〉 第二十章: 博學之, 審問之, 慎思之, 明辨之,篤行之。
107. John F. Krafcik, "Triumph of the Lean production system," *Sloan Management Review* 30, No. 1 (1988): 41–50.

108. The original Chinese text is as follows.〈論語〉里仁篇: 君子喻于義, 小人喻于利。
109. Abraham H. Maslow, *Motivation and Personality* (Harper Collins, 1987).
110. The original Chinese text is as follows.〈論語〉顏淵篇: 子貢問曰, 有一言而可以終身行之者夫? 子曰, 其恕手。己所不欲, 勿施於人。
111. The original Chinese text is as follows. 司馬遷〈前赤壁賦〉: 客亦知水與月夫? 逝者如斯, 而未嘗往也。盈虛者如彼, 而卒莫消長也。蓋將自其變者而觀之, 則天地曾不能以一瞬, 自其不變者而觀之, 則物與我皆無盡也。
112. The original Chinese text is as follows.〈老子〉第五十八章: 禍兮福所倚, 福兮禍所伏。
113. Wikipedia, "Sima Qian" https://en.m.wikipedia.org/wiki/Sima_Qian (Accessed on 08/08/2017)..
114. Joseph E. Stiglitz, "Essay on Wealth Disparity," *South China Morning Post*, December 31, 2016.
115. Wikipedia, "Silicon Valley" https://en.m.wikipedia.org/wiki/Silicon_Valley (Accessed on 12/08/2017).
116. *Wall Street Journal*, July 22–23, 2006.
117. Christophe Lecuyer, *Making Silicon Valley: Innovation and the Growth of High Tech 1930–1970* (The MIT Press, 2007).
118. Chong-Moon Lee, William F. Miller, Marguerite Gong Hancock and Henry S. Rowen, Co-editors, *The Silicon Valley Edge: A Habitat for Innovation and Entrepreneurship* (Stanford: Stanford University Press, 2000).
119. Christophe Lecuyer, "Fairchild Semiconductor and its Influence," in Ref. 118 (2000).
120. "Silicon Valley: To Fly, To Fall, To Fly Again," *The Economist*, July 25, 2015.
121. Henry S. Rowen, "Serendipity or Strategy: How Technology and Markets came to Favor Silicon Valley," in Ref. 118 (2000).
122. Annalee Saxenian, "Networks of Immigrant Entrepreneurs," in Ref. 118 (2000).
123. M. H. Xing, "Experiences of Economic Development of Taiwan: Legacy of the Positive and the Negative," in Charles H.C. Kao and Joseph S. Lee, Editors, *The Taiwan Experience: Past and Future*. (Taipei: Commonwealth Publishing), 1995. The original Chinese text is as follows. 邢慕寰,"台灣經濟發展的經驗: 正面與反面的教訓" 高希、李誠主編 "台灣經驗再定位" 天下文化出版, 台北, 1995.
124. Lawrence J. Lau, "The Economy of Taiwan, 1981-1988: A Time of Passages", in Lawrence J. Lau, Editor, *Models of Development*, ICS Press, 1990.
125. Wikipedia, "Hsinchu Science and Industrial Park" https://en.m.wikipedia.org/wiki/Hsinchu_Science_and_Industrial_Park (Accessed on 08/08/2017).

126. Wang Shi-Chi, *Interviewer and Editor: The Father and Son Team of Tsinghua: Presidents Shu Shien Shieu and Frank Hsia-sen Shu* (Hsin-Chu: National Tsing-Hua University Press, 2012). The original Chinese text is as follows. 王仕琦 採訪撰稿: "父子双傑清華傳承: 徐賢修與徐遐生兩位校長的故事". 新竹: 國立清華大學出版社, 2012.
127. Hong Yi-Yen, *ITRI- Engine of Innovation*: *The Mover of Taiwan's Industrial Success* (Taipei: Commonwealth Publishing, 2003). The original Chinese text is as follows. 洪懿妍, "創新引擎工研院: 台灣產業成功的推手" 台北: 天下雜誌, 2003.
128. The Group of Advisors changed over time. Serving the longest are the following: Frederick Seitz (President, NAS), Norman Hackerman (President, Rice U), Pierre Aigrain (Minister Sci-Tech, France), Umberto Columbo (Minister-Energy, Italy), Edward McKay (VP-Western Electric), Ivan Bennett (VP, NYU), Robert Evans (VP, IBM).
129. Poh-Kam Wong and Chee-Yung Ng, Editors. *Industrial Policy, Innovation and Economic Growth: The Experience of Japan and the Asian NIEs* (Singapore University Press, 2001): Chapters 12 & 13.
130. Zengzi said: "I examine myself daily on three aspects- Have I not committed loyalty to my job? Have I not cultivated trust with my friends? Have I not practiced what I have learned?" The original Chinese text is as follows. 〈論語、學而篇〉: 曾子曰: 吾日三省吾身, 為人謀而不忠乎? 與朋友交而不信乎? 傳不習乎?
131. Richard C. Paddox, "Family dispute over house of Singapore's founder erupts as national crisis," *The New York Times*, July 5, 2017. Also, on the same subject, Ashok Kumar Murpuri, *The New York Times*, July 12, 2017.
132. Hu An-Kan, *China*: *Innovation and Green Development*, (Beijing: China Ren-Min University, 2012). The original Chinese text is as follows. 胡鞍鋼, "中國: 創新绿色發展", 中國人民大學出版社, 北京, 2012.
133. "Leadership changes at Giant Bicycles," *Commercial Times*, December 16, 2016. Text in Chinese.
134. EPISTAR, http://www.epistar.com.tw/ (Accessed on 07/08/2017).
135. Ardentec, http://www.ardentec.com (Accessed on 07/08/2017).
136. Bruce Cheng, Narrator. *Solid Power-The Business Philosophy of Bruce Cheng and Delta Electronics* (Taipei: Commonwealth Publishing, 2012).
137. "Sales of McDonald's China," *South China Morning Post*, January 16, 2017.
138. "Jack Ma Engages in Business Diplomacy with Trump," *South China Morning Post*, January 10, 2017.
139. "Alibaba Group- Internal Speeches of Ma Yuen: Takes Hope for Tomorrow," The Hon-Qi Publishers, 2015. The original Chinese text is as follows. 阿里巴巴集團, "馬雲內部講話: 相信明天", 紅旗出版社, 2015.

140. "Alibaba is largest Tax Payer in China," *South China Morning Post*, January 4, 2016.
141. Michael Schuman, "A Table Maker Fights The Fakes on Alibaba," *The New York Times*, March 19, 2017.
142. Jan Cai, "Jack Ma Set Up Scholarship in Memory of Old Friend," *South China Morning Post*, February 4, 2017.
143. Wu Xiao-Bo, http://dushu.qq.com (Retrieved on 07/08/2017).
144. Jason L. Q Zeng, Personal Interview, Hong Kong, October 12, 2016.
145. Wikipedia, "Tencent" https://en.m.wikipedia.org/wiki/Tencent (Accessed on 07/08/2017).
146. Ying-Li Liu, Personal Interview, Shenzhen, 2017.
147. Wu Xiao-Bo, *Tencent Corporation: The Evolution of an Internet Company in China from 1998-2016*, Zhejiang University Press, 2017. The original Chinese text is as follows. 吳曉波 (2017) "騰訊傳: 1998–2016 中國互聯網公司進化論"。浙江大學出版社。
148. Sun Cheng, The Road to Knowledge Economy in Taiwan, (Taipei: San-Min Books, 2001). The original Chinese text is as follows. 孫震: "台灣發展知識經濟之路"。台北 三民書局, 2001.
149. Chintay Shih, Kung Wang and Yi-Ling Wei, "Hsinchu, Taiwan-Asia's Pioneering High Tech Park", in Henry S. Rowen, Marguerite Gong Hancock and William F. Miller, Co-Editors, *Making IT; The Rise of Asia in High Tech* (Stanford University Press, 2007).
150. Benoit Godin, "National Innovation System: The System Approach in Historical Perspective," Project on the History and Sociology of DTI Statistics. Working Paper No. 36 (2007), http://www.csiic.ca (Accessed on 07/08/2017).
151. Willy Shih, Chen-Fu Chien, Chintay Shih and Ting-Cheng Chen, "System on a Chip 2008: Ardentec Corporation," *Harvard Business School Case Study*, N9-609-026. August 25, 2008.
152. Dennis Wong, "The iPhone: A Sum of Parts," *South China Morning Post*, September 12, 2016.
153. Susan Strasser, "What's in your Microwave Oven?" *New York Times*, April 15, 2017.
154. Texas Instrument Announcement, *PR Newswire*, April 4, 2017.
155. Suzanne Burger, Richard K. Lester and the MIT Made by Hong Kong Study Team (Oxford University Press, 1997).
156. David Dodwell, "It's time to bridge the Gap," *South China Morning Post*, November 28, 2015.
157. Niki Sun and Gary Cheung, "Hong Kong's 12th Fiscal Surplus to Fund Education, Innovation and Salary Rebates," *South China Morning Post*, February 20, 2017.

158. P. L. Heng, *Neither Civil nor Servant: The Philip Yeo Story* (Singapore: Straits Times Press, 2016).
159. Andrew Liptak, "China's first automated cargo spaceship has docked with the Tiangong 2 space station," *The Verge*, April 22, 2017 http://www.theverge.com/2017/4/22/15393328/china-first-automated-cargo-spaceship-tianzhou-1-tiangong-2-space-station (Accessed on 07/08/2017).
160. Eugene Tang, "China lifts bar in high-speed rail with new technology," *South China Morning Post*, September 26, 2016.
161. Wikipedia, "Asian Infrastructure Investment Bank" https;//en.m.wikipedia.org/wiki/Asian_Infrastructure_Investment_Bank (Accessed on 07/08/2017).
162. Wang Zhenmin, as quoted by David Dodwell, "The Voice of Authority," *South China Morning Post*, April 16, 2016.
163. Joint Study of Chinese Academy of Engineering and the Hong Kong Academy of Engineering Sciences, May 24, 2017.
164. Joseph Needham, "Science and Civilization of China," Vol. 2, *History of Scientific Thought* (Cambridge University Press, 1956); Corrected Reprint, Cambridge University Press, 1991.
165. Michael D. Shear, "Trump Will Withdraw U.S. from Paris Climate Agreement'. *The New York Times*, June 1, 2017.
166. Oliver Milman, "G20 Leaders' Statement on Climate Change Highlights Rift with U. S." *The Guardian.com*, July 8, 2017.
167. Sam Costello, "How Many iPhones have been Sold World-wide". *Hotwire.com*, July 1, 2017.
168. Otto. C. C. Lin, "A Matter of Trust- What Both Sides of the Occupy Divide Can Learn from Confucius," *South China Morning Post*, October 16, 2014. *SCMP.com*, October 15, 2014.
169. "Big Mac Index," *The Economist*, July 13, 2017
170. Bill Vlasic, "G. M. Profits Falls Sharply as its Challenges Mount," *The New York Times*, July 26, 2017.
171. Jack Ewing, "Across Europe, Diesel's Future Appears Shaky," *The New York Times*, July 26, 2017.
172. Li Yuan, "Alibaba, Tencent Must Tread Cautiously," *Wall Street Journal*, June 30, 2017.
173. Elizabeth Dwoskin, Craig Timberg and Brian Fung, "Tech Industry Opposition to Trump Travel Ban Escalates," *The Washington Post*, February 2, 2017.
174. Arista Balakrishnan and Harriet Taylor, "Silicon Valley Leaders Slam Trump Travel Plan, Offer Employees Legal Aid," *Cnbc.com*, (Accessed on 28/01/2017).
175. Jack Ewing, "VW Loses Research Funding After Cheating on Emissions," *The New York Times*, August 2, 2017.

176. Raymond Cheng, "Incentive Needed to Drive Bay Area Plan," *South China Morning Post*, June 21, 2017.
177. Amanda Lee, "Industry Titans Endorse Greater Bay Area Vision," *South China Morning Post*, June 21, 2017.
178. Andrew Higgins, "The Private Business Sector Clashes with State Security," *The New York Times*, August 10, 2017.
179. Julian Burger and Justin McCurry, "Donald Trump Vows to answer North Korea nuclear threats with Fire and Furry". *The Guardian*, 9 August, 2017. Also, Justin McCurry, "Donald Trump on North Korea: All Options Are on the Table", *The Guardian*, 29 August, 2017.
180. https://www.policyaddress.gov.hk/2017/eng/speech.html(Accessed25/10/2017).
181. Jun Mai "Graft Fight Given Bigger Footing, Confident Xi's 'New Era' for China", *South China Morning Post*, October 19, 2017.
182. Simon Denyer," China's Xi Jinping unveils his top party leaders". *The Guardian*, October 25, 2017.
183. "Kobe Steel admits falsifying data on 20,000 tonnes of metal," *The Economist*, October 12, 2017.
184. Sean McLain and Chieko Tsuneoka, "Kobe Steel admits 500 companies misled in scandal," *Wall Street Journal*, 13 October 2017.
185. Leo Lewis, "Being the Boss of scandal-hit Kobe Steel is no bed of roses," *FT Weekend*, October 14–15, 2017.
186. Cheng Yu-Jie, "The Great Guangdong-Hong Kong-Macau Bay Area: Strength, Weakness and Opportunity," *Ten Yung*, June, 2017. The original text in Chinese as follows. 鄭宇部,"粤港澳大灣區：起点、痛点與奇点", 騰雲 060, 2017.
187. Nectar Gan, "Xi Ally is new Guang-dong Party Chief," *South China Morning Post*, October 29, 2017.
188. Peggy Sito and Maggie Zhang, "HK Slips to 12th in Survey of Most Competitive Cities," *South China Morning Post*, October 31, 2017.
189. The Global Urban Competitiveness Report 2017–2018. http://www.gucp.org/GUCP/Main/English (Accessed 08/11/2017).
190. Climate Science Special Report - 4th National Climate Assessment (NCA4) https://science2017.globalchange.gov/chapter/executive-summary/ (Accessed 06/11/2017).
191. Ni Yi-Qing and Xiao-Wei Ye, Editors, "Proceedings of the 1st International Workshop on High-Speed and Intercity Railways," (Springer-Verlag, 2012).

APPENDIX

List of Acronyms & Definitions

A*STAR	—	Agency for Science, Technology and Research
AIIB	—	Asian Infrastructure Investment Bank
AlGaInP	—	LEDs composed of aluminium-gallium-indium-phosphite
B2B	—	Business to Business
B2C	—	Business to Customer
BARI	—	Belt and Road Initiatives
BMRC	—	Biomedical Research Council
BRICS	—	Brazil, Russia, India, China and South Africa
C2C	—	Customer to Customer
CAE	—	Chinese Academy of Engineering
CAGR	—	Compound Annual Growth Rate
CAS	—	Chinese Academy of Sciences
CCL	—	Computer and Communication Laboratories
CEO	—	Chief Executive Officer
CISOH	—	Center of Industrial Safety and Occupational Health
CMOS	—	Complementary metal-oxide-semiconductor
CNRC	—	China National Railway Corporation
CoCom	—	Coordinating Committee for Multilateral Export Controls
CSIRO	—	Commonwealth Scientific and Industrial Research Organization

CSRD	—	China Southwest Jiao-tong University Railway Development Corporation
CTO	—	Chief Technology Officer
DRAM	—	Dynamic Random-access Memory
DSI	—	Data Storage Institute
DSO	—	Defence Science Organization
EDB	—	Economic Development Board
EMI	—	Electromagnetic interference
ERSO	—	Electronic Research and Service Organization
EV	—	Electric Vehicle
FDIC	—	Federal Deposit Insurance Commission
FhG	—	The Fraunhofer Society
GaAs	—	Gallium Arsenide
GaP	—	Gallium Phosphite
GDP	—	Gross Domestic Product
GinTech	—	Grumman International-Nanyang Technological Institute for CAD/CAM
HBLED	—	High Brightness Light Emitting Diodes
HKAES	—	Hong Kong Academy of Engineering Sciences
HKSE	—	Hong Kong Stock Exchange
HSIP	—	Hsinchu Science-based Industrial Park
HSMT	—	Health and Safety Monitoring Technology
HSR	—	High-speed Rail
I2R	—	Institute for Infocomm Research
IaaS	—	Infrastructure as a Service
IBN	—	Institute of Bioengineering and Nanotechnology
IC	—	Integrated Circuit
IDM	—	Integrated Device Manufacturer
IHPC	—	Institute of High Performance Computing
IMCB	—	Institute of Molecular and Cell Biology
IME	—	Institute of Microelectronics
IMRE	—	Institute of Materials Research and Engineering

InGaN — Indium Gallium Nitride
IoE — Internet of Everything
IPR — Intellectual Property Rights
I&T — Innovation and Technology
IRU — Industrial Research Unit
ISP — Internet Service Provider
ISS — Institute of System Science
ITI — Information Technology Institute
ITIC — Industrial Technology Investment Corporation
ITRI — Industrial Technology Research Institute
IVA — Royal Swedish Academy of Engineering Sciences
JAIST — Japan Institute of Advanced Science and Technology
JTC — Jurong Town Corporation
JV — Joint Venture
KIST — Korea Institute of Science and Technology
LAMDA — Look, Ask, Model, Discuss and Act
LBNL — Lawrence Berkeley National Laboratory
LCD — Liquid Crystal Display
LED — Light-emitting Diode
MBE — Molecular Beam Epitaxy
MIRL — Mechanical Industry Research Laboratories
MOCVD — Molecular organic chemical vapor deposition
MOEA — Ministry of Economic Affairs
MRL — Materials Research Laboratories
NAE — National Academy of Engineering
NAM — National Academy of Medicine
NAS — National Academy of Science
NASA — National Aeronautics and Space Administration
NCA4 — Fourth National Climate Change Assessment
NGO — Non-government Organization
NIH — National Institute of Health
NIST — National Institute of Science and Technology

NL	—	National Laboratories
nm	—	nanometer
NRC	—	National Research Council
NRF	—	National Research Foundation
NSC	—	National Science Council
NSF	—	National Science Foundation
NSTB	—	National Science and Technology Board
OECD	—	Organization for Economic Cooperation and Development
OEM	—	Original Equipment Manufacturer
OES	—	Opto-Electronics and System Laboratories
PaaS	—	Platform as a Service
PARC	—	Palo Alto Research Center
PRD	—	Pearl River Delta
PROMIS	—	A computer control program of manufacturing information system
R&D	—	Research and Development
RCA	—	Radio Corporation of America
RDAS	—	Research and Development Assistance Scheme
RFIC	—	Radio Frequency Integrated Circuit
RIEC	—	Research, Innovation and Enterprise Council
ROC	—	Republic of China
SaaS	—	Service as a Service
SAHTECH	—	Safety and Health Technology Center
SAR	—	Special Administrative Region
SARS	—	Severe Acute Respiratory Syndrome
SEC	—	Securities and Exchange Commission
SERC	—	Science and Engineering Research Council
SEZ	—	Special Economic Zone
SIMTech	—	Singapore Institute of Manufacturing Technology
SISIR	—	Singapore Institute of Standards and Industrial Research
SLAC	—	Stanford Linear Accelerator Center
SME	—	Small and Medium-sized Enterprise

SOC	—	System on a Chip
SOE	—	State-owned Enterprises
SPRING	—	Standards, Productivity, Research and Innovation Board
SRAM	—	Static Random-access Memory
STAG	—	Science and Technology Advisory Group
SWOT	—	Strength Weakness Opportunity Threat
TEEMA		Taiwan Electrical and Electronic Manufacturers Association
TNO	—	The Netherlands Organization for Applied Scientific Research
UCL	—	Union Chemical Laboratories
ULSI	—	Ultra-Large-Scale Integrated Circuits
UMC	—	United Microelectronic Corporation
USPTO	—	United States Patent and Trademark Office
VLSI	—	Very-Large-Scale Integrated Circuits
WTO	—	World Trade Organization

List of Companies/Organizations

3Com
Acer
Adobe
Advanced Micro Devices (AMD)
Advanced Packaging Technology Co., Ltd (APT Corporation)
Airbnb
Alcatel-Lucent
Alstom
Alibaba Group / Taobao.com / TMall.com
Amazon
Amelco
American Telephone and Telegraph Company (AT&T)
America Online (AOL)
Apple Computer (Apple Inc.)

Ardentec
Amana Corporation
Asus
ASM
Autonetics

Babcock & Wilcox
Baidu
Beckman Instruments
Bellcore Corporation
Bell Laboratories (Bell Labs)
Biosensors
Bombardier
Boeing Corporation
BP
BYD

Capital Management Service/Sequoia Capital
Carlyle Group
China International Marine Containers
China Merchants Bank
China Mobile Communications Corporation (China Mobile)
China Motors Group
Cisco Systems
CITIC Capital
Colgate-Palmolive
Compaq
Creative Technology

Davis and Rock
DEC
Dell
Delta Electronics
DHL
Digital
DJI
Dropbox
DSO National Laboratories

Eastman Kodak Company
eBay
E.I. du Pont de Nemours (Du Pont)
Electroglas
Energy Laboratory
Energy and Resources Laboratory (ERL)
Epistar
Ericsson
Exxon

Facebook
Fairchild Camera and Instrument
Fairchild Semiconductor
Federal Express (FedEx)
Fidelity
Ford Motor Company
Foxconn
Fujitsu

GeLumed
Genentech
General Dynamics
General Electric
General Instruments
General Micro Electronics
General Motors (GM)
GIANT Manufacturing Co.
GlaxoSmithKline
Glaxo-Welcome
Goldman Sachs
Google

Hewlett-Packard
Hitachi
High Speed Rail Corp
Hon Hai Corporation of Taiwan
Huawei
Hughes
HTC

IBM
ICQ
Infineon
Intel
International Data Group (IDG)
Intesils

Johnson and Johnson Company

Kawasaki Heavy Industries
Kasper
Keynote
Kleiner Perkins Caufield and Byers (KPCB)
Kobe Steel
Konka

Lilly
LinkedIn
Lockheed Corporation (Lockheed-Martin)
Lotus Engineering
LSI Logic
Lucent
Luminus

McDonald's
Merck Sharp and Dohme
Microsoft
MITRE Corporation
Motorola
Multichoice International Holdings (MIH) Group

National Semiconductor
NetEase
Netflix
Netscape
Nokia Bell Labs

Oracle
Oxford Committee for Famine Relief (Oxfam)

Pacific Century Cyber Works (PCCW)
Panasonic
Philips Electronics
Philips Petroleum
Ping An Bank
PR Newswire
Proctor and Gamble

Qian Zhao Electronics, Inc.

Raytheon Corporation
Rheem Semiconductor
Roche
Rockwell

San'an Optoelectronics Co., Ltd
Samsung
Seagate
SGS
SGS-Thomson (now STMicroelectronics)
Shell
Shockley Semiconductor Laboratory
Siemens
Signetics
Sina
Skype
Skyworth
Soft Bank (Japan)
Sohu
Sperry
Sperry-Univac
SRI (Stanford Research Institute) International
Starbucks
Sumitomo Chemicals
Sun Microsystems

Tappan Company
Taiwan Electrical and Electronic Manufacturers Association (TEEMA)
Taiwan Semiconductor Manufacturing Corp (TSMC)

Takeda
Tencent Holdings Ltd. (Tencent)
Texas Instruments Incorporated (TI)
TP Link
Twitter

Uber
United Parcel Service (UPS)

Vanguard International Semiconductor Manufacturing Company (VIS)
Vanke
Varian Associates
Volkswagen AG

Walmart
Wang Laboratory
WeChat
Western Electric
Wright-Patterson Air Force Base

Xerox Corporation (Xerox)

Yahoo!
YouTube
Yue-Loong Motors

ZTE

List of Universities

Aalborg University
City University of Hong Kong (CityU)
Chinese University of Hong Kong (CUHK)
Chinese Academy of Railway Sciences
Cambridge University
California Institute of Technology (Caltech)
Columbia University
Fudan University
Harvard University

Hangzhou Normal College
Hong Kong University of Science and Technology (HKUST)
Massachusetts Institute of Technology (MIT)
Nanjing University
National Cheng-kung University (NCKU)
National Jiao-tong University (NJTU)
National Taiwan University (NTU)
National Tsing-hua University (NTHU)
Oxford University
Peking University
Shenzhen University
South China University of Technology
Southern University of Science and Technology (SUSTech)
Southwest Jiao-tong University
Stanford University
The Hong Kong Polytechnic University (HKPolyU)
Tsinghua University
University of California, Berkeley (UC-Berkeley)
University of California, San Francisco (UC-San Francisco)
University of Hong Kong (HKU)
University of Michigan
University of Sussex
Xi'an Jiao-tong University
Zhejiang University

List of Personalities

Abraham Maslow
Ada Wen-shung Ma (馬文松)
Alexander Bell
Alex P.K. Wai (衛炳江)
Alfred Coyle
Alfred Cho (卓以和)
Alfred Sloan

Amancio Ortega
Andrew Carnegie
Andrew Grove
Anthony Ku (谷家恆)
Arthur Rock

B-A Lundvall
Barack Obama
Bill Gates
B. J. Lee (李秉傑)
Brian Chesky
Bruce Cheng (鄭崇華)

Carrie Y.N. Lam-Cheng
Carlos Slim Helu
C. C. Hang (康長傑)
Chao-yang Zhang
C. Freeman
Chenye Xu (Daniel) (許晨曄)
Chiang Ching-kuo(蔣經國)
Chiang Kai-shek (蔣介石)
Chi Ming Chang (張季明)
Chi Mo Wang (汪其模)
Chintay Shih (史欽泰)
C.P. Chan (陳中屏)
Confucius (孔夫子)
Cornelius Vanderbilt
C. Y. Lu (盧志遠)

Ding Lei
Daniel Y.C. Ng (伍日超)
Deng Xiao-ping (鄧小平)
D.H. Hu (胡定華)
D. M. Jiang (姜達銘)
Donald Trump
Donald Valentine

Eduardo Saverin
Elvis Deng (鄧茂松)

Eric Lean (林耕華)
Eugene Kleiner

Franklin D. Roosevelt
Frederick Terman
Fred Lee

Garrett Camp
Gilbert Chin (陳瑞礎)
Gordon Moore
Guan-chao Zhou (周光召)
Guo-tang Zhao (趙國堂)

Harry Gatos
H.C. Fang (方賢齊)
Hellen Keller
Henry Ford
Henry Ying-tong Fok (霍英東)
Herbert Hoover
Hiroya Kawasaki
Henry S. Rowen
Hsuan Yeh (葉玄)
H.Y. Tam (譚華耀)

J.A. Schumpeter
Joseph Tsai
Jacob Chang-lin Tarn (譚晶琳)
Jennifer Hou (侯玉貞)
Jay Last
J.D. Rockefeller
Jean Hoerni
Jeff Bezos
Jerry Yang
Joe Gebbia
John Bardeen
Joseph Tsai (蔡崇信)
Jun Wu
Johnsee Lee (李鍾熙)
Julius Blank

Jyo-ming Shyu (徐爵民)

K. K. Lee
K. B. Liu (劉金標)
K.T. Li (李國鼎)
Kao-wen Mao (毛高文)

Larry Ellison
Laozi (老子)
Lee Kuan-yew (李光耀)
L.Y. Chen (陳陵援)
Lionel Ni (倪明選)
Liqing Zeng (Jason) (曾李青)
Ludwig van Beethoven

Ma Huateng (Pony Ma) (馬化騰)
Mao Zedong (毛澤東)
Mark Zuckerberg
Ming Seng Guan
Ma Yun (Jack Ma) (馬雲)
Mencius (孟子)
Ming Keon Teo
Michael Bloomberg
Ming-jun Chou (周銘俊)
M.S. Lin
Ming Hseung Lin (林敏雄)
Morris Chang (張忠謀)

Otto Chui-chau Lin (林垂宙)

Pang Xiang Wang (王鵬翔)
Philip Yeo (楊烈國)
Pierre Du Pont

Richard C. Liu (劉中鴻)
Roger J.X. Liu (劉榮憲)
Robert Langer
Reynold Liu (劉榮隆)
Robert H.C. Tsao (曹興成)
Robin Li

Robert Noyce
Robert H. S. Chen (陳興時)
Robert J. C Yang (楊日昌)
Rodrigo Duterte
R.R. Nelson

Sergey Brim
Sheldon Roberts
Shirley S.L. Chang
Sima Qian (司馬遷)
S.L. Ho (何紹鎏)
S.S. Shu (徐賢修)
Stephen W. L. Tsai (蔡維崙)
Steve Jobs
Steve R. Y. Cheng (鄭瑞雨)
Sun Yat-sen (孫逸仙)

Thomas Edison
Tony Tan
Tony Lo (羅祥安)
Travis Kalanick
Tze-Kung (子貢)
Tze-Lu (子路)

Victor Grinich
Vinod Khosla

Walt Disney
Walter Bratton
Warren Buffett
Wilbur and Orville Wright
William B. Shockley
W.Y. Pan (潘文淵)

Xi Jinping (習近平)
Xiao Long Chin (秦曉隆)
Xu-sen Lin (林樹森)

Y.C. Liu
Yidan Chen (Charles) (陳一丹)

Ying-li Liu (劉應力)
Yiqing Ni (倪一清)
Y.S. Sun (孫運璿)

Zhi-tung Wang
Zengzi (曾子)
Zhu Rongji (朱鎔基)
Zidong Zhang (Tony) (張志東)
Zhong Ming Liu (劉仲明)

PHOTOGRAPHS

Meeting with Premier Chiang Ching-kuo. Taipei, August, 1977.

Meeting at MOEA on National Development Conference. Taipei, August, 1977.

Otto Lin at Inauguration of ITRI Presidency, 1988, Taipei, with Morris Chang and Li-An Chen (Minister of Economic Affairs) officiating.

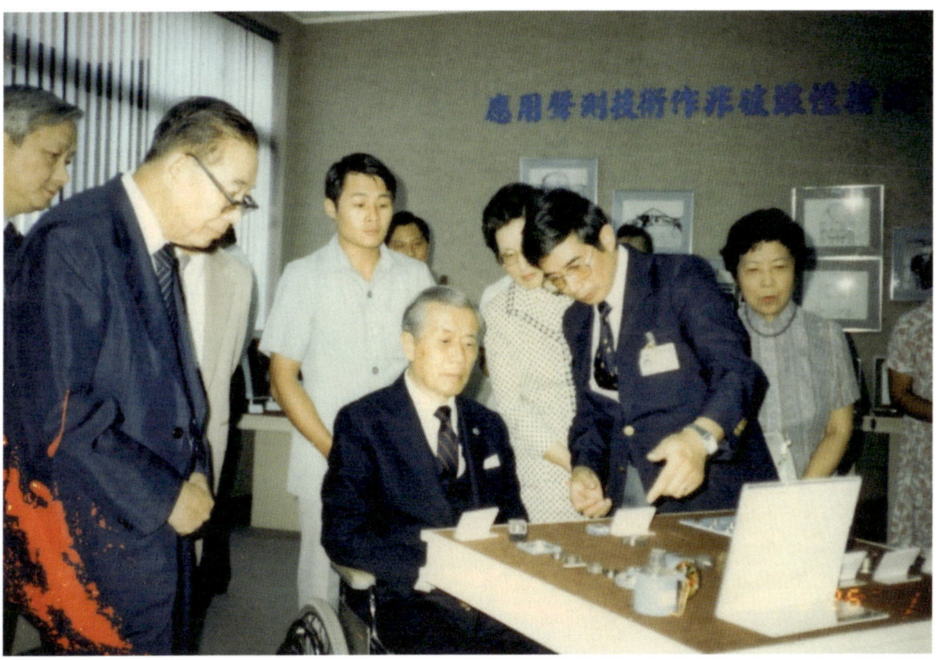

At Materials Research Laboratories (MRL)/ITRI, Hsinchu, 1986.

Back: Morris Chang, (back center), Mrs. Y.S. Sun, Mrs. H.C. Fan Front: H.C. Fan, Premier Yun-Shuan Sun, Otto Lin.

Signing ceremony with GIANT Manufacturing Co. to develop carbon fiber composite bicycles. Taipei, 1985.
Second from left: K.B. Liu (GIANT), Otto Lin, M.S. Lin, Reynold Liu.

Dr. H.C. Fang and Dr. S.S. Shu at a contest, lifting strong rare-earth magnetic bars from a test board. Hsinchu, 1986.

Hosting dignitary visitors at ITRI. Hsinchu, 1991.
From left: Otto Lin, T.H. Lee (President), K.T. Li (Senior Minister), L.A. Chen (Minister), K.A. Hsu (Director of Industry).

Visiting Council for Scientific and Industrial Research (CSIR), South Africa. Pretoria, 1990. From left: Otto Lin (President, ITRI) and Chris Garber (President, CSIR).

Hosting Singapore Premier Goh Chok Tong at ITRI, 1991.

Addressing ITRI team on quality training at Boeing Corporation, Seattle, 1992.

Meeting of the Executive Yuan Science and Technology Advisory Group (STAG).
Front: Fred Seitz (U.S.), Pierre Aigrain (France), Umberto Columbo (Italy), X.Y. Lee (ROC), W. Starr (U.S.).
Back (from right): Otto Lin (4th), Y.S. Shih (5th), P.Z. Wu (6th). Lyons, 1991.

Visiting Dr. Max Sabre (President, The Fraunhofer Society). Munich, Germany, 1991.

Groundbreaking ceremony, ITRI-VLSI Lab, in HSIP, 1992.
From left: Y.S. Shih, Steve Cheng, Klaus Weimer, Morris Chang, Otto Lin, Chintay Shih, Nick Lu, C.T. Xing, X.H. Chen, John Lu.
This lab was to become the first fab line of TSMC.

Industrial Technology Research Institute (ITRI) Innovation Plaza, completed 1995, in Hsinchu.

Ceremony of Inauguration, October 24, 1992. The Royal Swedish Academy of Engineering Sciences (IVA), Stockholm.
From left: Alastair North (Asian Institute of Technology), Merton Miller (University of Chicago, Nobel Laureate in Economics, 1990), Endel Lippmaa (Estonian Academy of Sciences), Otto Lin (ITRI), Ulrich Seiffert (Volkswagen).

Hosting Dr. Alfred Cho, AT&T Bell Labs. Taipei, 1992.

Discussion with industrial leaders.
From right: Otto Lin (1st), T.H. Lee (President, 2nd) at ITRI, 1991.

Speaking at the National Assembly on Science, Technology and Industry. Yang Min-shan, Taipei, December 1992.

Group of four ITRI Presidents. 2012, Taipei.
From left: J.M. Shyu (4th), Otto Lin (1st), Chintay Shih (2nd), Johnsee Lee (3rd).

With Chia-wei Woo (President, HKUST) at ceremony. Clear Water Bay, Hong Kong, 1997.

At Congregation, HKUST, 2000.
From left: Professor C.N. Yang (Nobel Laureate in Physics, 1957) and Otto Lin.

With HKUST colleagues at a trip to the Silk Road, Urumqi, 1998.
From left: T.C. Pong, Roland Chin, Sabrina Cheng, Zena Chin, Otto Lin, Ada Lin, Ping Cheng, Ann Lin, Mrs. Pong, Helen Shen.

At Ceremony of Donation for Nansha Education Program-HKUST, 2005.
From left: Henry Ying-tong Fok, Otto Lin and Sir S.Y. Chung (back view).
Standing in back row: Paul C.W. Chu (President), Siu-ming Kwong.

A bird's eye view of the Nansha IT Park, Guangzhou, 2003.

Soft Opening-First Phase of Nansha IT Park, Nansha, Guangzhou, December 29, 2002.
Speaking at the podium: Otto Lin.
In the backgound of Dr. Sun Yet-sen's calligraphy: K.N. Zhang, K.C. Chen, X.Y. Zhou, Xu-sen Lin (May or of Guangzhou), Henry Fok, Yuen-he Lin.

Nansha Forum on Innovation. Nansha, Guangzhou, 2003.
From left: Lan Xue (Tsinghua University), Henry Rowen (Stanford University), Otto Lin (HKUST), Richard Nelson (Columbia University), Michael Kelly (University of Ottawa).

Global Innovation Forum at Salzburg, 2006.
From left: Philip Yeo (EDB/A*STAR-Singapore), Valentin Smyntanya (ONU, Ukraine), Otto and Ada Lin (HKUST), Jean and Judson King (UC-Berkeley), Taizuo Yakusuzi (Tokyo University, Japan).

Meeting with Premier Zhu Rongji. At Zhongnanhai, Beijing, March 25, 1993.

Opening Ceremony of Symposium in Materials Science.
From right: Premier Y.S. Sun, Professor Guan-chao Zhou (President, CAS), Professor T.Y. Wu (President, Academia Sinica), Professor H.M. Hsia (Chairman, NSC).
Speaking: Otto Lin. Taipei, January 1994.

At HKUST Forum on Commercialization of Research Results, 2003.
From left: Ying-li Liu (Vice Mayor of Shenzhen), Guan-hua Xu (Minister at Ministry of Science and Technology), Otto Lin (Vice President of R&D, HKUST).

At Tsing-Tech Innovations Ltd, Hong Kong, 2005.
Back row: Christopher Shih, Jack Lau, T. Yang, Siao-min Lee.
Front row: Herbert Lee, Otto Lin, Han-min Zheng, Quan-sen Zhang, Ada Lin.

Meeting with Beijing-Shanghai High Speed Rail Corp. 2011.
From left: Otto Lin, Timothy Tong (President, HKPolyU), Guo-tang Zhao (Vice President of Engineering, JF HSR), Alex Wai (VPRD-HKPolyU), T.L. Ng (Council Director, HKPolyU).

Test ride on Beijing-Shanghai HSR with colleagues of HKPolyU. Beijing South Station. June 2011.
Standing from left: Li Cheng (3rd), T.L. Ng (5th), Tim Tong (6th), Wang Hon (7th), G.T. Zhou (8th), Alex Wai (9th), Otto Lin (10th), P.X. Wang (11th), Michael Liu (12th).
Standing from right: Fred Ng (1st), Yiqing Ni (2nd).
Front row from right: Andrew Young (1st).

Lecturing at CNRS — North Train Corporation at Chang Chun, 2015.
From left: Jin Tien Wang (Vice President) and Otto Lin.
In the background, wall displaying institutional partnership.

Attending Tencent Conference on Greater Guangdong-Hong Kong-Macau Bay Area. June 2017, Hong Kong. From right: Chenggang Xu, Lawrence J. Lau, Wen Hai, Fanny Law, Otto Lin, Guanghan Chen.

Lecturing at Southern University of Science and Technology with Professor Ke Liu, addressing questions from audience. Shenzhen, October 2016.

Otto Lin and Guan-chao Zhou (President, CAS) meeting with the press. Taipei, 1994.

Otto Lin meeting students at an outreach conference at HKUST, Hong Kong, 1999.

In signing ceremony with The Netherlands Organization for Applied Scientific Research (TNO), Taipei, 1992.

Otto Lin with S.S. Shu (Former Chairman, ITRI) at ITRI"s 20th Anniversary.